Brackish water can be found wherever sea water mixes with fresh water. The abundant estuaries and coastal lagoons of northwestern Europe are therefore brackish-water habitats, as are a variety of smaller, and often manmade, coastal ponds. Such environments possess a rich fauna, as witnessed by their use by birds and fish as feeding and/or nursery areas. Many are also under threat from rising sea levels, the construction of sea defences, coastal land reclamation and drainage, and pollution.

This book examines and reviews the ecology of the brackish-water animals and habitats of northwestern Europe, and forms the first complete identification guide to the fauna of these ecologically important areas. The keys are specifically designed to be used with living material, and illustrations are provided for all the species, making identification easy. Field workers, naturalists and students interested in Europe's coastal fringe will find this an indispensable reference.

The brackish-water fauna of northwestern Europe

The brackish-water fauna of northwestern Europe

An identification guide to brackish-water habitats, ecology
and macrofauna for field workers, naturalists and students

R.S.K. BARNES

Department of Zoology and St Catharine's College,
University of Cambridge, UK

CAMBRIDGE
UNIVERSITY PRESS

Published by the Press Syndicate of the University of Cambridge
The Pitt Building, Trumpington Street, Cambridge CB2 1RP
40 West 20th Street, New York, NY 10011-4211, USA
10 Stamford Road, Oakleigh, Melbourne 3166, Australia

First published in 1994

Printed in Great Britain at the University Press, Cambridge

A catalogue record for this book is available from the British Library

Library of Congress cataloguing in publication data

Barnes, R. S. K. (Richard Stephen Kent)
The brackish-water fauna of northwestern Europe: a guide to brackish-water habitats, ecology, and
macrofauna for field workers, naturalists, and students/R.S.K. Barnes.
    p.   cm.
Includes bibliographical references (p.   ) and index.
ISBN 0 521 45529 4 (hc). – ISBN 0 521 45556 1 (pb)
1. Brackish water fauna – Europe.   2. Brackish water fauna – Europe – Indentification.
3. Brackish water ecology – Europe.   I. Title.
QL253.B37   1994
591.94–dc20   93-39843 CIP

ISBN 0 521 45529 4 hardback
ISBN 0 521 45556 1 paperback

For Morvan Keri and Annaëlle Gwenllyn

Contents

Preface   xi

Sources of illustrations   xiii

PART I Brackish-water biology

Introduction   1

The origin of brackish-water habitats   6

Brackish-water habitats and their investigation   16

Nature of brackish-water faunas: estuaries   27

Nature of brackish-water faunas: lagoons   30

The diets of brackish-water animals   33

What limits the populations of brackish-water animals?   36

Reproduction   45

PART 2 Identification and notes on the species

Identification   49

Key to major groups of brackish-water animals   52

Cnidarians   56

Flatworms and nemertines   70

Nematodes   76

Annelids   78

Sipunculans   104

Molluscs   105

Crustaceans   154

Insects   206

Bryozoans   244

Echinoderms   250

Ascidians   252

Fish   254

References   272

Index to organisms   281

Preface

I remember, several years ago, listening regularly to *Round Britain Quiz*, a radio programme which posed such esoteric questions as: 'What have the shooting of birds, the treatment of certain wingless insects, and books on brackish-water animals in common?'. The answer was always obvious once it has been explained.

I have on my bookshelves the authoritative 2400-page *Synopsis and Classification of Living Organisms* published by McGraw-Hill in 1982. If one seeks to look up in that book the Diplura, Protura and Collembola – three groups of small wingless insects – one uncovers an interesting situation. The introduction to the account of the Insecta indicates that the three groups above are not in fact insects and infers that they will therefore be covered in that section of the treatise that is devoted to myriapods. On turning to that section, however, it is obvious that its authors, like many others, consider that they *are* insects, and so they are not covered there either. Alone amongst groups of living organisms then known, the Diplura, Protura and Collembola were omitted from that otherwise encyclopaedic work because no-one wished to lay claim to them.

Brackish-water animals have suffered from the same kind of problem. Fresh water biologists have assumed that they fall within the province of the marine biologist, and *vice versa*. Moreover, different groups of animals or different types of brackish-water have traditionally been the concern of people from different disciplinary backgrounds: the crustaceans have been studied mainly by marine biologists, for example, but the insects by land-based workers; estuaries have generally been approached from the sea, but lagoons appear to have been approached largely by accident. The end result of this is that although there are numerous guides to the faunas of fresh and marine waters, respectively, there is no single work which will permit the identification of animals from brackish waters. Anyone foolish enough specifically to be interested in this type of habitat has had to possess a small library of books and specialist papers otherwise devoted to elements of the fresh water, marine and even terrestrial faunas. Hence the production of this book.

And so to the second motivating force. Not all that many years ago, if a naturalist saw a bird that he or she could not immediately name it would have been entirely acceptable for him or her to shoot the bird in order to identify it. If that were done in northwest Europe today, however, there would be an outcry and the action could even lead to prosecution. The ethos of today is: 'better that the bird should live unidentified than for it to be killed simply in order to put a name on it'. Yet, if the unknown animal is a crustacean or an insect or a mollusc, then it seems that nothing has changed since the days when the essential possession of a naturalist was the means of killing. 'Kill first and identify afterwards' seems the creed of most invertebrate field workers. I for one cannot grasp the logic whereby a bird's life must be safeguarded whereas a prawn's can be sacrificed, and I believe and hope that within a few years concern for nature will extend beyond the land vertebrates and abstractions such as rain forests to the lives of individual invertebrates. Accordingly, the keys in this book have been designed for use without harm on living animals that can then be returned to their habitats after identification. Indeed, in some cases the characters concerned can *only* be seen whilst the animal is alive. I have therefore attempted to produce a guide that is equivalent to those widely available for amphibians, reptiles, birds and mammals, and I trust that this book will thereby not only facilitate study of what is still a neglected but widespread habitat type in Europe (as elsewhere), but that it will do so without necessitating the destruction of part of that which is being investigated.

Finally, I wish that I could claim that my coverage of the brackish-water macrofauna is complete, but I cannot. Many of the smaller brackish waters and their faunas remain uninvestigated, and taxonomic problems and difficulties of identification render some records in the literature open to question and information on several species patchy to say the least. Our knowledge of the fauna is therefore still incomplete; my coverage of estuarine and lagoonal species, however, is as full as I have been able to make it. A start has to be made somewhere, and doubtless future works will be able to improve considerably on this one. I should also point out that lagoonal habitats may appear to have been granted a disproportionate amount of attention in some sections of the introductory chapters. This is deliberate. Within the general category of brackish waters, the estuarine habitat is the best known and most intensively researched. Nevertheless, semi-isolated, non-tidal, pond-like brackish waters are abundant in north-western Europe and are more easily accessible than the majority of estuaries, even if very few words have been devoted to them in the existing popular and technical literature. I have endeavoured to redress the balance somewhat and to draw the attention of a wider audience to what we do know of their ecology and evolution.

I am most grateful to Drs Paul Cornelius, Laurie Friday and Peter Hayward for valuable advice and assistance in respect of various, to me obscure, hydroids, beetles and bryozoans, respectively; to Dr Colin Little for adding species to my original draft list; to Annaïg Darmorie and Marc Weller for translating various troublesome passages from the European literature into English; and to Dr Sally Corbet for suggestions for rendering the introductory sections into equally intelligible English. I would also like to record my indebtedness to Adolf Remane, Wilhelm Schäfer and Bent Muus who, via their writings, have always nurtured my interest, and provided copious insight, into the lives of the animals treated in the following pages.

RSKB
May 1993

Sources of illustrations

With a handful of exceptions, I have based the illustrations in this book on those available in the literature, and I list the sources that I used in (re)drawing my figures below. A few have been reproduced, with permission, directly from previously published sources. The exceptions were drawn from life or from my own photographs.

Fig. 1   Salinity data after Head (1972), for the Tyne, and Kühl (1972) for the Elbe.

Fig. 2   After Reid & Wood (1976)

Fig. 3   From a photograph in the collection of the Committee for Aerial Photography, University of Cambridge

Fig. 4   After Green (1968)

Fig. 5   The 50 and 100 m depth contours after Lee & Ramster (1981)

Fig. 6   After Hardisty (1990)

Fig. 7   After Carter (1988)

Fig. 9   After Muus (1967)

Fig. 10 and 11   From photographs in the collection of the Committee for Aerial Photography, University of Cambridge

Fig. 12   After Nelson-Smith (1965)

Fig. 13   After Mangelsdorf (1967)

Fig. 14   After Head (1972)

Fig. 15   After Dorey *et al.* (1973)

Fig. 16   After Barnes *et al.* (1988)

Fig. 17   After Barnes (1989c) and the author's photographs

Fig. 18   After Barnes (1991)

Fig. 19   *Fucus ceranoides* after Hiscock (1979); *Lamprothamnium papulosum* after Moore (1986); others after Clapham *et al.* (1965)

Fig. 21   After Barth & Anthon (1956) and Lincoln (1979)

Fig. 22   After Fenchel & Kolding (1979)

Fig. 23   After Muus (1967)

Fig. 24   After Barnes (1994b)

Fig. 25   From data in Zwarts & Blomert (1992)

Fig. 26   *Hydrobia ulvae* after Fish & Fish (1977); others after Thorson (1946)

Fig. 28   *Protohydra leuckarti* after Muus (1967); all other polyps after Hayward & Ryland (1990); *Sarsia tubulosa* medusa after Russell (1953)

Fig. 29   *Cordylophora caspia* and *Clava multicornis* after Hayward & Ryland (1990); *Bimeria franciscana* after Morri (1981); *Clavopsella navis* after Millard (1975); *Rathkea octopunctata* after Russell (1953); *Halitholus cirratus* – polyp after Forsman (1972), medusa after Ackefors & Hernroth (1972)

Fig. 30   All after Hayward & Ryland (1990)

Fig. 31   *Obelia bidentata* and *Hartlaubella gelatinosa* after Hayward & Ryland (1990); *Obelia* medusa after Russell (1953); *Aurelia aurita* after Barth & Anthon (1956); anemones after Manuel (1988)

Fig. 32   *Cereus pedunculatus* and *Edwardsia ivelli* after Manuel (1988); the others after Hayward & Ryland (1990)

Fig. 34   *Procerodes littoralis*, *Uteriporus vulgaris* and *Dugesia lugubris/polychroa* after Ball & Reynoldson (1981); *Leptoplana tremellaris* after Prudhoe (1982); and *Cephalothrix rufifrons* and *C. linearis* after Gibson (1982)

Fig. 35   *Lineus* spp. from Gibson 1982, with permission; *Amphiporus lactifloreus* after the same author

Fig. 36   *Prostomatella obscurum* after Brunberg (1964); nematode after Fitter & Manuel (1986); all others after Gibson (1982)

Fig. 38   *Aphrodita aculeata*, *Harmothoe imbricata* and *Lepidonotus squamatus* after Fish & Fish (1989); *Harmothoe spinifera* and *Pholoe minuta* after Fauvel (1923); *H. sarsi* after Forsman (1972); *Gattyana cirrhosa* after Hartmann-Schröder (1971)

Fig. 39   *Phyllodoce groenlandica* after Fauchald (1977); *P. maculata*, *Eteone longa*, *E. picta* and *Eulalia viridis* after Pleijel & Dales (1991); phyllodocid from McIntosh (1873–1923); *Glycera* after Fish & Fish (1989); *Exogone naidina* after Rasmussen (1973) and *Streptosyllis websteri* after Hartmann-Schröder (1971)

Fig. 40   *Perinereis* and *Nephtys* from McIntosh (1873–1923); parapodia all after Fauvel (1923)

Fig. 41   *Scoloplos armiger* from McIntosh (1873–1923); *Malacoceros fuliginosus* and *Scolelepis squamata* after Fish & Fish (1989); *Paraonis fulgens* after Hartmann-Schröder (1971); *Streblospio shrubsoli* after Hartmann-Schröder (1971) and Forsman (1972); *Scolelepis foliosa* and *Spiophanes bombyx* after Fauvel (1927); *Marenzelleria viridis* after Maciolek (1984)

Fig. 42   *Polydora ligni*, *P. pulchra* and *P. ligerica* after Hartmann-Schröder (1971); *P. ciliata* from McIntosh (1873-1923); *Magelona mirabilis* after Hayward & Ryland (1990); *Pygospio elegans* after Ushakov (1955); *Spio filicornis* after Fauvel (1927)

Fig. 43   *Tharyx marioni* and *Caulleriella zetlandica* after Fauvel (1927); *Cirratulus cirratus* and *Cirriformia tentaculata* from McIntosh (1873–1923); *Psammodrilus balanoglossoides* after Westheide (1990); *Capitella capitata* and *Arenicola marina* after Fish & Fish (1989); *Heteromastus filiformis* after Hartmann-Schröder (1971)

Fig. 44   *Armandia*, *Melinna palmata* and *Ampharete grubei* after Fauvel (1927); *Ophelia bicornis* after Harris (1980); *O. rathkei* after Hartmann-Schröder (1971); tubes of *Sabellaria spinulosa* after Schäfer (1972); *S. alveolata* reef from a photograph by the author; *Alkmaria romijni* after Thorson (1946)

Fig. 45   *Amphitrite figulus* after Hartmann-Schröder (1971); tube of *Lanice conchilega* from McIntosh (1873–1923); *Terebellides stroemi* after Fauchald (1977); *Fabricia stellata*, *Sabella pavonina* and *Myxicola infundibulum* after Fish & Fish (1989); *Fabriciola baltica* after Hayward & Ryland (1990); *Manayunkia aestuarina* after Fauvel (1927)

Fig. 46   *Ficopotamus enigmatica* and spirorbid tubes after Bianchi (1981); lumbriculid and *Helobdella stagnalis* after Quigley (1977); naidid, tubificid and enchytraeid after Hayward & Ryland (1990); *Nephasoma minuta* after Gibbs (1977)

Fig. 48   *Leptochiton asellus* and *Lepidochitona cinereus* after Jones & Baxter (1987); the others after Graham (1988)

Fig. 49   All after Graham (1988)

Fig. 50   *L. vincta* after Graham (1988) and Fretter & Graham (1962); *Littorina saxatatilis*,*L. saxatilis* var. *tenebrosa*, *L. littorea* and *L. obtusata* after Graham (1988); *L. saxatilis* var. *lagunae* after Muus (1967); and *L. mariae* after Hayward & Ryland (1990)

Fig. 51   *Hydrobia ulvae*, *H. neglecta* and *H. ventrosa* after Muus (1963 and 1967); *Potamopyrgus antipodarum* after Graham (1988) and Fretter & Graham (1962); *Heleobia stagnorum* after Bank & Butot (1984) and Anon; and *Pseudamnicola confusa* after Graham (1988)

Fig. 52   *Truncatella subcylindrica* after Graham (1988) and Fretter & Graham (1962); *Rissoa parva* after Hayward & Ryland (1990); all others after Graham (1988)

Fig. 53   *Rissostomia membranacea/labiosa*, *Paludinella littorina* and *Rissoella globularis* after Graham (1988); *Assiminea grayana* after Graham (1988) and Fretter & Graham (1962); *Rissoella diaphana* after Hayward & Ryland (1990); and *R. opalina* after Fretter & Graham (1962) and Hayward & Ryland (1990)

Fig. 54   All after Graham (1988), except *Caecum armoricum* which is after Fretter & Graham (1962)

Fig. 55   *Brachystomia eulimoides* after Graham (1988) and Fretter & Graham (1962); *B. rissoides* and *Turbonilla lactea* after Graham (1988); *Diaphana minuta, Retusa obtusa* and *R. truncatula* after Thompson (1988)

Fig. 57   *Berthella plumula* after Hayward & Ryland (1990); others all after Thompson (1988)

Fig. 58   All after Thompson (1988)

Fig. 59   *Doridella batava* after IUCN (1983); *Goniodoris nodosa* after Hayward & Ryland (1990); others all after Thompson (1988)

Fig. 60   *Tenellia adspersa* and *Eubranchus farrani* after Hayward & Ryland (1990); others after Thompson (1988)

Fig. 62   *Lymnaea pereger* after Janus (1965); *Crassostrea gigas* original; all others after Hayward & Ryland (1990)

Fig. 63   *Astarte borealis* after Barth & Anthon (1956); all others after Hayward & Ryland (1990)

Fig. 64   *Cerastoderma glaucum* after Petersen (1958); *Parvicardium hauniense* after Petersen & Russell (1971a); all others after Hayward & Ryland (1990)

Fig. 65   *Ensis directus* after van Urk (1987); all others after Hayward & Ryland (1990)

Fig. 66   *Mya arenaria* after Yonge & Thompson (1976) and Barth & Anthon (1956); *Congeria cochleata* after Nyst (1835); *Mya truncata* after Barth & Anthon (1956) and Hayward & Ryland (1990); *Corbula gibba* and *Barnea candida* after Hayward & Ryland (1990)

Fig. 67   Ostracod after Athersuch *et al.* (1989); barnacles all after Hayward & Ryland (1990)

Fig. 69 and 70   All after Tattersall & Tattersall (1951)

Fig. 72   *Diastylis rathkei* after Jones (1976); *Cyathura carinata* after Richardson (1905); *Eurydice pulchra* after Forsman (1972); others after Naylor (1972)

Fig. 73   *Idotea* spp. and *Asellus aquaticus* after Forsman (1972); others after Naylor (1972)

Fig. 75   *Apseudes latreillii, Tanais dulongii* and *Heterotanais oerstedi* after Holdich & Jones (1983); *Lysianassa ceratina, Ampelisca brevicornis* and *Gitana sarsi* after Lincoln (1979)

Fig. 76   *Talitrus saltator* after Hayward & Ryland (1990); all others after Lincoln (1979)

Fig. 77 and 78   All after Lincoln (1979)

Fig. 79   *Bathyporeia pilosa* after Forsman (1972); all others after Lincoln (1979)

Fig. 80   *Pontoporeia femorata* after Forsman (1972); *Pontocrates* spp. after Lincoln (1979) and Chevreux & Fage (1925); all others after Lincoln (1979)

Fig. 81   *Gammarellus angulosus* after Sars (1895); *Atylus swammerdami* after Hayward & Ryland (1990); all others after Lincoln (1979)

Fig. 82   All after Lincoln (1979)

Fig. 83   *Corophium* after Barth & Anthon (1956); *Corophium insidiosum* after Hayward & Ryland (1990) and Lincoln (1979); *C. curvispinum* and *C. acutum* after Lincoln (1979); all others after Hayward & Ryland (1990)

Fig. 84   All after Lincoln (1979)

Fig. 86   All after Smaldon (1979) and Hayward & Ryland (1990)

Fig. 87   *Crangon crangon* after Smaldon (1979); all others after Christiansen (1969)

Fig. 90   *Anurida maritima* and *Petrobius brevistylis* after Hayward & Ryland (1990); *Cloeon dipterum* nymph after Clegg (1956); *Ischnura elegans* nymph after Fitter & Manuel (1986); *Sympetrum sanguineum* and *Orthetrum cancellatum* nymphs after Hammond (1983)

Fig. 91   Corixid after Green (1968); fore leg and pala of *Sigara striata* after Jansson (1986); all other palae after Savage (1989)

Fig. 92   *Hesperocorixa sahlbergi, Notonecta viridis* and *Gerris* sp. after Savage (1989); *Corixa affinis* and *C. panzeri* after Jansson (1986); *Saldula palustris* after Green (1968); *Notonecta glauca* after Palmer (1993)

Fig. 93   *Limnephilus* and *Oecetis ochracea* larvae after Wallace *et al.* (1990); *Mystacides longicornis* larva after Hickin (1967); case of *Acentropus niveus* after Clegg (1956); *Nymphula nympheata* caterpillar after Fitter & Manuel (1986); chironomid larva after Macan (1959)

Fig. 94   Dolichopodid, ptychopterid, tabanid and ephydrid larvae after Merritt & Cummins (1984); psychodidae larva after Smith (1989); stratiomyiid and syrphid larvae after Macan (1959); and tipulid larva after Clegg (1956)

Fig. 95   *Gyrinus marinus, Haliplus apicalis, H. obliquus* and *H. confinis* after Holmen (1987); *G. caspius* after Joy (1932); *Gyrinus* larva after Macan (1959); *Haliplus* larva and *Haliplus* after Richoux (1982); and *H. wehnckei* after Holmen (1987) and Friday (1988)

Fig. 96   *Haliplus immaculatus* after Holmen (1987) and Friday (1988); *H. flavicollis* after Holmen (1987); *Noterus clavicornis* after Richoux (1982); *Noterus* and *Colymbetes* larvae after Klausnitzer (1991); *Colymbetes fuscus* after Green (1968)

Fig. 97   *Rhantus exsoletus* after Richoux (1982); *Rhantus* and *Agabus* larvae after Klausnitzer (1991); *Agabus biguttatus* after Joy (1932); *A. bipustulatus* after Harde (1981); *A. nebulosus* after Friday (1988); *A. conspersus* original drawing

Fig. 98   *Dytiscus* larva after Klausnitzer (1991); *Coelambus impressopunctatus* after Friday (1988); all others after Richoux (1982)

Fig. 99   *Hydroporus, Hydroporus* larva and *Potamonectes depressus* after Richoux (1982); *Potamonectes* larva after Klausnitzer (1991); *Hydroporus angustatus* after Joy (1932); *H. tessellatus* after Friday (1988); *H. planus* and *H. pubescens* original drawings

Fig. 100   *Helophorus minutus* and *Helophorus* larva after Hansen (1987); *H. flavipes* after Harde (1981); other *Helophorus* spp. after Friday (1988); *Berosus* after Richoux (1982); *B. affinis* and *B. spinosus* after Friday (1988); *Berosus* larva after Macan (1959)

Fig. 101   *Enochrus bicolor* and *E. halophilus* after Hansen (1987); *E. melanocephalus* after Friday (1988); *Laccobius biguttatus, Paracymus aeneus* and larvae of *Enochrus, Laccobius* and *Paracymus* after Richoux (1982)

Fig. 102   *Hydrobius fuscipes* after Richoux (1982); *Hydrobius* larva after Macan (1959); *Ochthebius marinus* after Hansen (1987) and Friday (1988); *O. nanus* original drawing; all other *Ochthebius* spp. after Friday (1988)

Fig. 103   *Macroplea mutica*, adult and larva after Richoux (1982); *Bledius spectabilis* after Green (1968); *Heterocerus* larva after Klausnitzer (1978); staphylinid larva after Merritt & Cummins (1984)

Fig. 104   *Heterocerus fenestratus* and *Cillenus laterale* after Green (1968); *H. obsoletus, H. flexuosus* and *H. maritimus* after Clarke (1973); *Cillenus* larva after Reitter (1908)

Fig. 106   *Plumatella repens* and *P. fungosa* after Occhipinti Ambrogi (1981); all others after Hayward (1985)

Fig. 107   *Bowerbankia imbricata, Conopeum seurati, C. reticulum* and *Electra crustulenta* after Ryland & Hayward (1977); others after Occhipinti Ambrogi (1981)

Fig. 108   *Callopora aurita* and *Turbicellepora avicularis* after Hayward & Ryland (1979); *Bugula* spp. after Occhipinti Ambrogi (1981) and Ryland & Hayward (1977); *Cryptosula pallasiana* after Occhipinti Ambrogi (1981); *Amphipholis squamata* after Barrett & Yonge (1958)

Fig. 109   *Botryllus schlosseri, Styela clava* and *Molgula manhattensis* after Millar (1970); *Petromyzon marinus* and *Lampetra fluviatilis* after Wheeler (1969)

Fig. 111–117 All after Wheeler (1969)

PART I Brackish-water biology

Introduction

There are four major types of aquatic habitat on this planet. Two of these – fresh waters and the sea – are much studied, relatively well known and form the subject of many a book. The third variety comprises inland salt lakes or seas: bodies of salt water that have derived their saltiness not from the sea but from salts dissolved out of their watersheds by the inflowing rivers. It is not often appreciated that these cover as much land area as fresh waters. Most such 'athalassic' systems occupy depressions in the continental land surface. Rivers flow in but not out, and evaporation is the process roughly balancing water input. The very small quantities of salts borne in by the rivers each year have therefore been concentrated over time until now many of these water bodies are extremely salty, even to the extent that salts may crystallize out around their margin. The larger ones have salinities intermediate between those of sea water and fresh water, and are to that extent brackish according to the letter of scientific definitions of that word ('any water which exhibits salinity intermediate between sea water and fresh water' - Lincoln *et al.* , 1982; Whittow, 1984). Williams (1981) and Comin & Northcote (1990) provide reviews of these interesting systems.

The fourth category, 'brackish water' in both the intention and letter of technical definitions (i.e. where the salts are contributed by water from the sea), is formed along coastlines by mixing wherever sea water is diluted by fresh water to form water of intermediate salinity. In warm parts of the world brackish-water salinities may not be intermediate all the time, however, because, as in the athalassic waters above, during the hot season evaporation may raise the salinity from below that of sea water to well above it. Nevertheless, their salts are always derived from the adjacent sea. Three main types of brackish water are recognized (Remane & Schlieper, 1971): (1) estuaries, i.e. the regions through which rivers discharge to the sea (Fig. 1); (2) coastal lagoons, i.e. bodies of coastal water that are separated from an adjacent sea by barriers of sand or shingle but that nevertheless receive a significant input of sea water (the barriers usually take the form of chains of extremely elongate offshore islands parallel to the coast, Fig. 2, or longshore spits and barrier beaches, Fig. 3); and (3) inland seas such as the Baltic that have a very limited exchange with the ocean and into which copious fresh water discharges (Fig. 4). Other less important categories of brackish water include small high-level rock pools on rocky shores and salt pans on saltmarshes, and numerous man-made brackish habitats such as those drainage ditches that have faulty sluices, boating lakes fed by pumped sea water, scrapes in low-lying coastal bird reserves, and even a few of the Norfolk broads (for example Hickling Broad).

No tidy human system of classification adequately reflects the rich and continuously varying reality of nature. Distinct oceans, continents, ecosystems, animal populations,

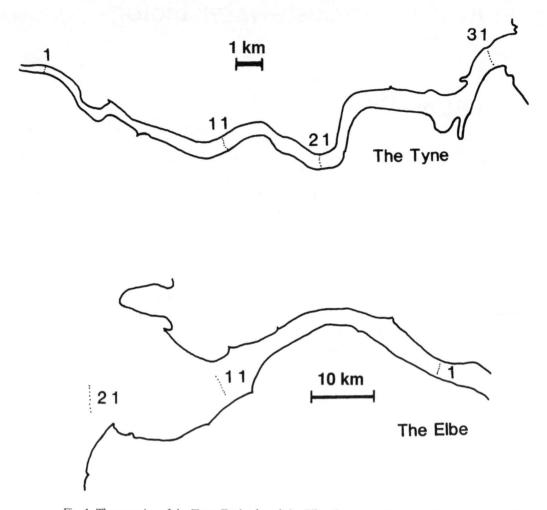

Fig. 1. The estuaries of the Tyne, England, and the Elbe, Germany. Typical surface salinities (in ‰) are indicated; the Elbe has a large freshwater discharge, the Tyne a smaller seawards flow.

etc., etc. do not actually occur in nature: they are really only figments of the human desire to put things into neatly labelled boxes and exist only by definition. Brackish waters are no exception. Thus, 'estuaries' merge insensibly into 'coastal lagoons' and both do so into the coastal sea and local fresh water environments; and 'inland seas' similarly merge into athalassic salt lakes in that many of the latter (such as the Caspian Sea) have, for historical and biogeographical reasons, a large brackish-water element in their faunas and floras dating back to times when they were genuinely brackish and connected to the ocean (Zenkevitch, 1963).

Because of their intimate connection with both fresh water and marine systems, brackish waters often contain a number of hardy fresh water and marine species – those that can withstand a degree of concentration and dilution, respectively, of the water in which they live. Indeed, it can be argued (Barnes, 1989a) that there are no

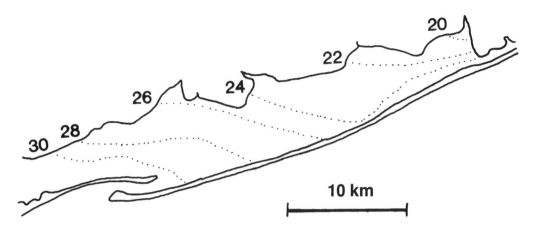

Fig. 2. A barrier-island lagoon: Great South Bay, Long Island, New York, USA. Salinities are indicated in ‰.

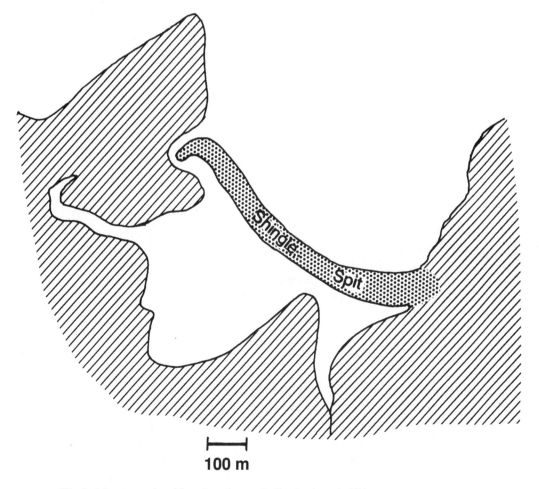

Fig. 3. A lagoon enclosed by a longshore spit: Cemlyn lagoon, Wales.

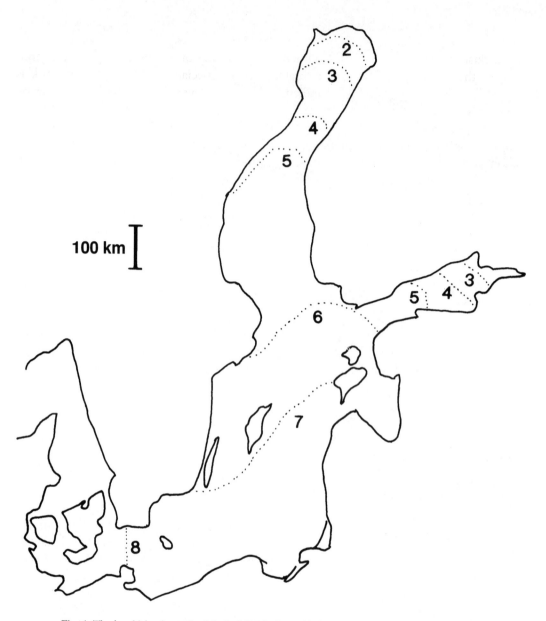

Fig. 4. The brackish, almost land-locked Baltic Sea, with its average surface salinity (in ‰)

truly brackish-water species in that all are capable of inhabiting one or other of the two larger habitat systems under the right conditions. Mediterranean authors collectively term brackish-water habitats the 'paralic system' to stress the point that their faunas and floras are not in fact restricted to brackish-waters although many of the species do not normally occur in the adjacent marine or fresh water habitats. One practical difficulty of this state of affairs is that it is impossible to produce a handbook covering all animals that might be found in brackish water: it would have to include vast numbers of essentially fresh water and coastal marine species. Fortunately, for a variety of reasons (explored below) a considerable number of species are much more characteristic of brackish waters or paralic environments than of any other habitat type. This handbook is therefore devoted to these 'brackish-water specialists'. Although this book is essentially addressed to a British audience, the area covered is northwestern Europe (the Atlantic coasts of Britain, Ireland and northern France, together with all English Channel coastlines and the fringes of the North Sea as far east as the Skagerrak, and as far north as the Shetlands and Bergen in Norway). This is a reasonably well-defined region and it has the merit of possessing a fairly uniform flora and fauna, although many brackish-water species have a much wider range, some being found right across the northern hemisphere including the Aral Sea. The Mediterranean Sea to the south and the Baltic Sea to the east have been excluded, however, because whilst they do contain many of the species described in this handbook each possesses a number of distinctive species unknown from northwestern Europe as defined above. No inland sea or athalassic water body occurs in the area covered, and hence this handbook will be devoted to estuaries, lagoons and similar habitats, and semantic problems of the word brackish can be ignored.

Amongst naturalists, these habitats are the particular haunt of those bird watchers interested in shorebirds and wildfowl, and although it is sometimes possible to see the food items being taken by the birds through a telescope, perhaps the majority of ornithologists completely ignore the invertebrates and small fish that feed the objects of their passion. The latter may not possess the grace and beauty of a curlew, avocet or shelduck, nor are they as obvious to the casual observer, but the prey species lead just as fascinating lives and in many cases these can be investigated much more easily than can those of the birds. The common lugworm, to take one familiar and widespread example, by causing water to flow through its burrow creates a special microenvironment favouring the growth of its own food species. In effect it 'gardens'. And in so doing it alters the nature of the habitat, to the advantage of some species and to the detriment of others. Whilst inoffensively lying in its burrow creating water currents, it may also have to defend itself against the activities of ragworms, which will nip its tail and seek to drive it from its home (see below). On rare occasions, lugworms may be observed swimming, in a somewhat ungainly manner, away from such sources of pestering. Their reactions to the presence of other species and the effect of (for example) the salinity of the water on their various activities can be observed in narrow aquaria such as can be constructed with a piece of rubber tubing in a U between two plates of glass. If the space within the U is filled with natural sediment overlain by aerated brackish water, many species will construct their burrow systems as if they were in the wild, and if the distance between the plates of glass (i.e. the thickness of the tubing) is set appropriately for the size of animal, the glass will form the side walls of the burrow and the animal inside will be fully visible. This is the principle on which much wildlife photography is based. Even the small mudsnails that crawl across the surfaces of brackish mudflats in densities of tens or hundreds of thousands per square metre have many a tale to tell, and indeed will do so if maintained in petri-dishes containing natural sediment and water. They are capable of distinguishing between different sediment types on the basis of particle size, food content, previous and present occupants, and salinity, for instance, and probably a whole range of other features. Simple either/or choice experiments can easily be set up in petri-dishes (one half of the dish containing sediment of type A and the other half of type B) and the preferences or powers of discrimination of the snails analysed. The effects of different environmental features on movement, on breeding and on a host of other activities can all be investigated in the same way. Fashionable brackish-water biology today involves complex mathematical models of how the whole or parts of the ecosystem works, but such models are only as good as the data upon which they are based, and all too often this is fragmentary (not least because it is unfashionable to study the lives of whole living organisms in the field). Yet it is relatively simple to reproduce segments of the habitat in the laboratory with 'string and sealing wax' and to observe the lives and interactions of animals in these microcosms. With a little ingenuity, much fundamental information can be gained – on reactions of individuals to increasing population density, on which species interfere with others, and on interactions between adults and juveniles, for example – and the efficacy of scientific modeling studies will thereby be enhanced.

The animals and habitats of today, however, are not just products of current processes and interactions. Much brackish-water biology will be a legacy of the evolutionary past, and throughout this introductory description of habitats and animals we must take into account the ways in which both environment and organisms originated and in which they have changed over the last few million years. This is especially important for brackish waters because, as we shall see, these types of habitat are ephemeral in time and are rarely static in space.

The origin of brackish-water habitats

All the present northwest European estuaries and lagoons were formed by the rise in sea level that followed the onset of the current interglacial period some 20 000 years ago. During the last two million years sea level has fluctuated, perhaps by as much as 200 m, dependent on how much water has been locked up as ice in glaciers and ice-caps. During the last 25 000 years, it has varied from being slightly higher than at present to being 150 m lower (Carter, 1988); 15 000 years before the present ('B.P.') sea level was about 100 m lower than it is now. During the last 20 000 years sea level has risen, encroaching on the land, most notably over what is now the southern North Sea (Fig. 5). Shallow areas of land were inundated with decelerating speed until the present position of the coastline was in general achieved some 4000 years B.P. (Fig. 6); over this period some one million square kilometres of land were lost. Although some of this inundation was achieved by erosion of soft cliffs, as is seen today along much of the North Sea coast of England, the primary route for the invasion of the sea would have been up river valleys. The lower parts of the then river valleys are now beneath the waves and what would once have been their upper/middle

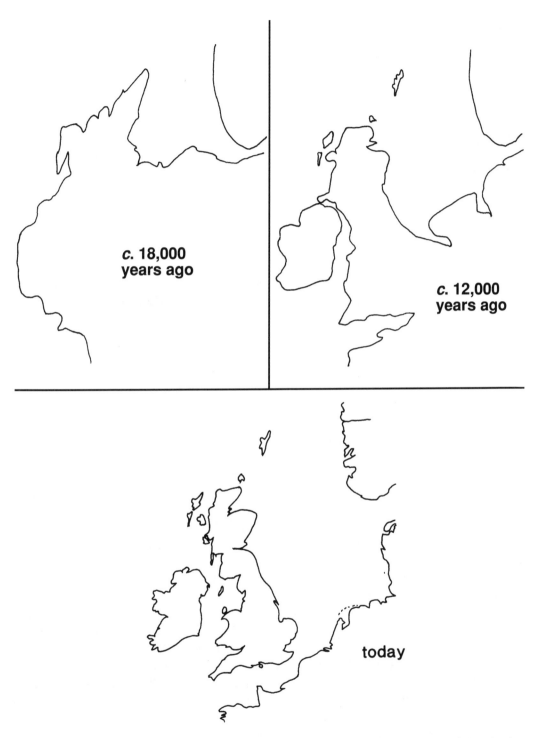

Fig. 5. Approximate changes to the coastline of the northwestern tip of Europe during the last 18 000 years. Based on curves such as that of Fig. 6 and the present position of the 100 and 50 m depth contours.

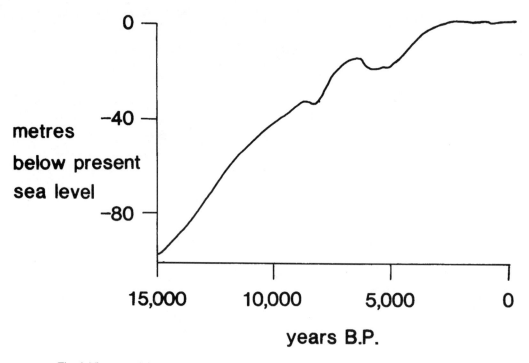

Fig. 6. The general form of the rise in sea level over the last 15 000 years.

reaches are the estuaries of today – the great majority of northwest European estuaries are drowned river valleys or 'rias'. In comparison with the rest of the world, northwest Europe is subject to large variations in tidal height (that is, it is a 'macrotidal' region with a height difference between high and low spring tides of more than 4 m; Davies, 1972). Because of this the invasion of the sea is re-enacted (here twice) every lunar cycle of 24 hours, 50 minutes. During high tide, sea water pulses in and immerses the beds of the estuaries, whilst during low tide the only water in sight may be the discharging fresh water of the river meandering in a small channel between extensive mudflats exposed to the air.

When ice covered the land, slowly moving glaciers scraped off and fragmented the surfaces of the underlying rock, eventually discharging much of this eroded material into the sea or spreading it over the land during glacial retreat. When sea level later rose, this material was pounded by waves and rolled by currents so that its angular surfaces were smoothed as individual pebbles were rotated one against another to form 'shingle' (technically particles larger than sands, comprising gravels (2–60 mm diameter) and cobbles (6–20 cm diameter)). The action of waves, especially during storms, was to push or roll this shingle into elongate masses parallel to the shoreline. Where the offshore regions were deep, these masses became plastered onto the coastline itself; where the offshore zone was shallow, the banks of pebbles formed 'offshore bars' at some distance away from the coast. And as the sea advanced towards and over the land, so it pushed the shingle masses ahead of it (Fig. 7).

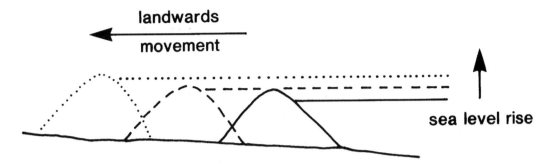

Fig. 7. Offshore shingle bars and their movement onshore as sea level rises.

Some of these shingle barriers diverted the mouths of estuaries, so that when an estuary had almost reached the sea its lower course became deflected through about 90° (Fig. 8). Other barriers, together with admixed sand and silt, partially blocked off marine bays, leaving only a very narrow channel through which tidal water could flow (the resultant high-water velocities serving to keep the channel open), and thereby formed lagoons (Fig. 9). Yet others sealed off the mouths of small estuaries completely, ponding back the discharging fresh water. Here, if the shingle barrier was low enough to be overtopped by high-tidal sea water, a brackish lagoon was again created (Fig. 10). More usually, however, the enclosing barriers were raised by storms to such heights that sea water could not overtop them and instead of a lagoon, a coastal fresh water lake formed.

Complete closure, as above, is not the only process preventing the evolution of partially enclosed marine bays into lagoons. By definition, a lagoon is a permanent body of water. But under a macrotidal regime, low tide often exposes extensive areas of intertidal zone, and during low tide most shallow semi-enclosed marine bays are largely exposed to the air; hence no lagoon forms. (Lagoons are most characteristic of 'microtidal' areas of the world – with spring tidal ranges of less than 2 m – where there is little tendency of the sea to move offshore during low tide.) What have developed in temperate macrotidal regions instead are extensive intertidal salt-marshes.

Especially around the North Sea barrier coasts, people have reclaimed these salt-marshes by using artificial isolating barriers – sea walls – to complete the process begun by nature. Nevertheless, unless it is further reclaimed for industrial development, any isolated salt-marsh or intertidal zone will remain below the level of high tide. Shingle is porous; water can percolate into a shingle barrier during high tide to seawards and it will there be joined by fresh water from rainfall to form a brackish water table within the barrier itself. This brackish water can then percolate out to landwards, onto the low-lying reclaimed land, and there fill any depressions remaining, the beds of former salt-marsh creeks, for example (Fig. 11), to form a distinctive type of lagoon (see Hunt, 1971, for the origin of another example). Surplus water then leaves such a lagoon via the drainage ditch systems that form an integral part of these reclamations, eventually to be discharged to the sea through sluices in sea walls. Most existing northwest European lagoons belong to this percolation-fed category, although their lagoonal nature may not be obvious. The shingle barriers may now be hidden beneath sand

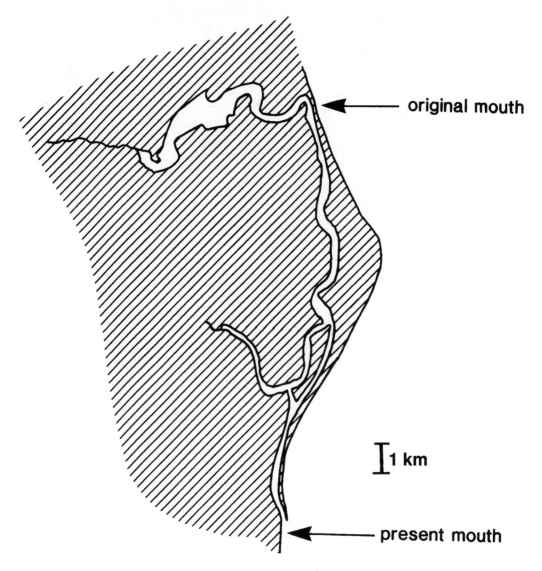

Fig. 8. A 15 km deflection of the mouth of the Alde/Ore Estuary, Suffolk, England, by the south-wards growth of the Orfordness shingle spit. During this deflection, the estuary has 'captured' that of the Butley River.

dunes, themselves almost obscured by afforestation, and the lagoon may look like a typical reed-fringed fresh water pond. Their salinities, however, can be a half to two-thirds strength sea water, and those lagoons that are more obviously associated with longshore shingle ridges usually contain almost full-strength sea water.

It is therefore evident from the above that whereas all estuaries must be brackish, at least during some stages of the tidal cycle, some lagoons may have salinities equal to that of sea water. But in the Atlantic climate of the northwestern tip of Europe, rainfall is plentiful (of the order of 1m per year) and rain is a potent source of diluting water

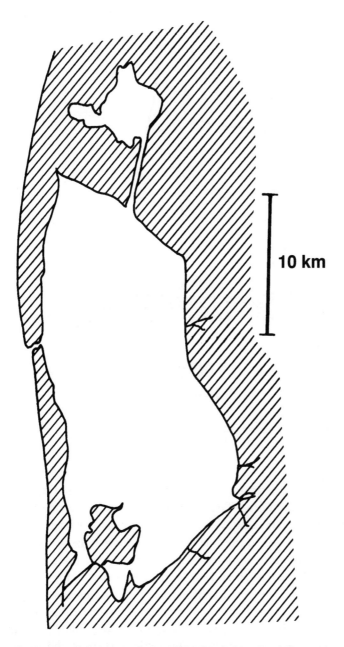

Fig. 9. A bar-built lagoon: the 30 000 ha Ringkøbing Fjord, Denmark.

for lagoons. Many also receive small streams and the groundwater discharges that out-flow at the coast. Thus, although from a worldwide perspective most lagoons are fully marine (Barnes, 1980), those of northwestern Europe do tend to be brackish.

Finally, people have created, usually unwittingly, many brackish ponds by excavating coastal pits into which some sea water can flow, via percolation or through faulty sluices

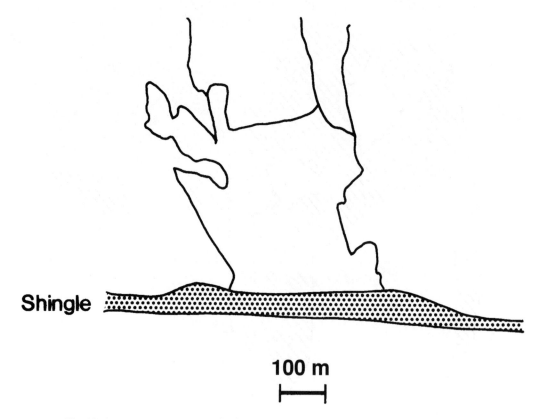

Shingle

100 m

Fig. 10. A completely land-locked lagoon: Benacre Broad, Suffolk, England. Sea water enters only by overtopping of the isolating barrier during some high tides.

(Barnes, 1989b). The drainage ditch systems of reclaimed land referred to above discharge to the sea through one-way sluices. These open or are opened during low tide but are designed to close at times of high tide. When their closure is imperfect – as a result of pieces of flotsam becoming lodged in the mechanism, or rusty hinges, or the wood of the sluice-flap becoming warped, etc. - some sea water will flow in during high tide. Salt water may then back up the drainage ditch and fill local depressions, being diluted *en route* by outflowing fresh water. Equivalent but usually deeper depressions have been created deliberately, for instance within onshore shingle barriers to extract pebbles for building, or to obtain the material with which to construct sea walls ('borrow pits'), and these are often filled by brackish water from the coastal water table, just as in the more natural cases mentioned above. Medieval fish ponds, more recent paddling pools, boating lakes and even ornamental moats around large houses are all further examples of artificially constructed depressions into which sea water may be admitted either accidentally or deliberately to form lagoon-like systems. Other excavations have been created behind sea walls by the force of the sea during the storm surges that affect the southern North Sea on average once every 25 years (see Wagret, 1959, or Pollard, 1978, for accounts of such surges). When high spring tides and gale force winds coincide, sea level can be much

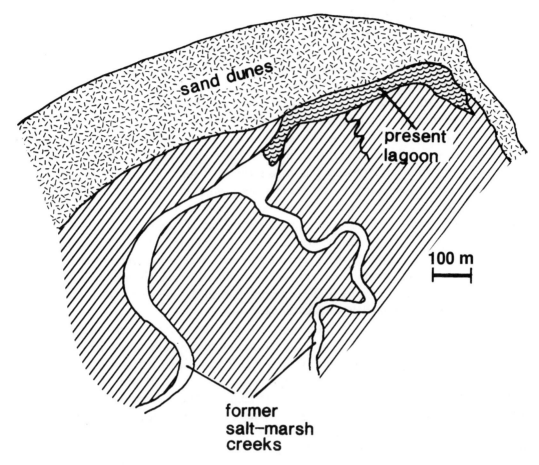

Fig. 11. A lagoon formed in part of the bed of a former salt-marsh creek: Broadwater, Norfolk, England.

higher than would be predicted from astronomical data, and such extreme high tides have often breached sea defences and flooded low lying coastal land. In The Netherlands in 1953, they caused the flooding of 143 000 ha and drowned almost 2000 people. Water rushed in with enough force to excavate pits up to 38 m deep (Ehlers, 1988). Other major inundations occurred in 1979 and 1993. The sea water that remains after repair of the breached sea wall may persist for many years, being progressively diluted by land drainage, and it may even be replenished by percolation.

There are many other ways in which artificial or semi-natural brackish ponds or lagoons have been formed; indeed, almost every one has its own singular history. One more example will help to give an impression of the diversity of their origins. As we have seen, shingle barriers may dam and pond back small rivers to form coastal fresh water lakes. The outflowing fresh water leaves such lakes by percolating seawards through the barrier. People have sometimes installed artificial inlet/outlet channels through the otherwise enclosing barriers for a variety of reasons – for instance to permit marine shellfish larvae or fish to enter, or to lower the lake level. The effect of constructing the channel has often been to permit sea water to enter at high tide, turning the former fresh water lake back into a brackish lagoon (Little *et al.*, 1973, describe the origin of one such).

Apart from the consequences of the Greenhouse effect and the isostatic effects of the removal of a great weight of ice from the land at the end of the last glacial phase (so that whilst central Scotland and southern Norway are still bobbing back up, as with a see-saw the southern North Sea and Western Approaches are subsiding – both at some 30 cm per century, as measured by tide gauges on the German North Sea coast), sea level now appears to have stabilized (Fig. 6). This stabilization has important repercussions on the fate of all brackish habitats. In effect, it means that their days are numbered. Estuaries tend to fill with sediment; a process that man seeks to counter by dredging in those estuaries that are used by shipping. Whilst sea level rises, an estuary is continually rejuvenated by being moved landwards, but once its position in space is fixed sediments will accumulate, the level of the bed will rise, and the estuary will revert to the river valley that it was before drowning. When the sea no longer rises, the shingle barriers that were pushed before the rising sea towards the ever-receding land will also finally reach the land. Throughout the world, offshore barriers, whether of sand or shingle, are getting closer to the shore. If they have not already been plastered onto the landwards high water mark itself, from which position they can be moved no further, they will soon be so, having moved over any lagoon that they once enclosed, permitting it to revert to the sea bed. 'Washover fans' – shingle pushed from the barrier into a lagoon by gale-whipped water – show this inexorable process at work. One study of a northern Atlantic lagoonal barrier system, in Rhode Island, U.S.A., has quantified the movement: 61% of the barrier sediment is being moved landwards, whilst the remaining 39% is being removed seawards (Fisher, 1980). Barriers are clearly not only moving landwards, they are getting thinner.

For reasons of coastal defence or shortage of agricultural or building land, people are seldom inclined to permit influx of sea water onto the land if it can be prevented without too great a cost. Barriers of shingle are regularly heightened by bulldozers in an attempt to prevent overtopping, thereby denying some lagoons their salt water input. As and when finances permit, sea walls are rendered less permeable, faulty sluices are replaced by more effective ones, and low-lying marshy areas are filled in because of the perception that they are 'wasteland'. Today, many existing brackish ponds survive only because they have been incorporated into nature reserves and the hydrological regime is managed to maintain them. Fortunately, it is relatively easy to create new lagoonal pools in areas between a shingle barrier and the true mainland high water mark; that is, in former salt-marshes that have been completely isolated from the sea by additional sea-wall barriers but that have not been infilled for industrial use. Most such areas were originally used for grazing. Any depression that can be excavated below the water table within the barrier is liable to fill with brackish or sea water. Such 'scrapes' have often been created in bird reserves. They are colonizable by brackish-water faunas and floras and serve to attract the waders and wildfowl that prey on them (for example at Minsmere in the U.K. – Axell & Hosking, 1977). Landwards movement of the shingle barriers will eventually obliterate these scrapes, however.

It is easy to recognise an estuary for what it is, although not all features on a map that look as if they are estuaries (i.e. like Fig. 12) may really be so. Unless significant

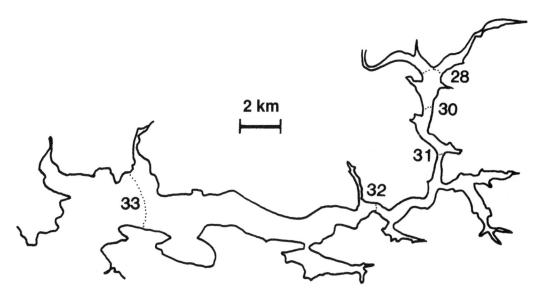

Fig. 12. Salinities (in ‰) within Milford Haven, Wales, during high water of spring tides: like several other drowned river valleys, for much of the time Milford Haven is virtually an arm of the sea and not a brackish-water habitat, although a number of estuaries do discharge into it.

rivers drain into a coastal inlet, it may be little more than an arm of the sea and only contain a minute, if any, brackish section. Several coastal features with the word 'Estuary' in their names are not in fact estuarine. (Conversely – although there is no example in northwest Europe – the river flow may be so great that the entire apparent estuary is fresh water: the mouth of the Amazon is one instance.) It should be clear from the above, however, that lagoons and brackish ponds may not be distinguishable from fresh water ponds at first sight; indeed many lagoons will eventually evolve into fresh water ponds. Any pond or lake that lies within 150 m or so of the sea may be brackish, especially if it is behind a sea wall or within or behind a shingle ridge. If the pond is brackish, then any drainage ditch issuing from it will also contain brackish-water organisms. If a pool is to landwards of a solitary sand-dune ridge, then it too may be brackish, particularly if there is evidence (most easily sought on the beach) that the dune has a shingle base, although lakes or pools within dune systems are less likely to be brackish. If the surrounding vegetation includes salt-marsh species then a marine influence is almost certain, but reeds (*Phragmites*) do not necessarily indicate fresh water status. If an instrument capable of measuring salinity (an appropriately calibrated hydrometer, refractometer or conductivity meter) is available, then again there is no difficulty: any reading above 1‰ (see below) indicates that the water is brackish. The most reliable indicator, however, is the nature of the fauna (see below), and if this is brackish or marine in character then it is well worthwhile walking around the pool to see if the means of sea water entry can be established. Are there any springs, for example, issuing into the pool along the margin nearest to the sea? Is the intervening barrier only just higher than the marine strand-line? Does water flow in from elsewhere along a drainage ditch running close to a sea wall? The manner of any fresh water entry can be identified at the same time. These features are not only interesting in their own right, but are also relevant to a host of ecological questions. For example – How might various elements of the fauna have gained access and is access still possible? What fluctuations in salinity and water level are likely within the habitat?

Brackish-water habitats and their investigation

Brackish waters provide four major types of habitats for animals:

(i)   the water itself;
(ii) expanses of soft sediment that are muddy and mainly exposed during low tide in estuaries, but are permanently submerged and often sandier and with gravel or shingle in lagoons and similar environments;
(iii) semi-aquatic stands of fringing vegetation; and
(iv) dense meadows of submerged water-weeds and algae, especially in the more pond- or lake-like systems.

By definition, the overriding feature of the water in brackish habitats is that its salinity lies between that of fresh water (by convention 0–1‰) and sea water (locally some 32–35‰; Lee & Ramster, 1981), where '‰' = parts per thousand by weight of dissolved salts (i.e. grams of salts dissolved in 1 kg of water). The salinity of estuaries must by their very nature vary from less than 1‰ at their head to that of the local sea water at or near their mouth (in the Kattegat and Belt Sea, however, the surface 'sea water' is itself brackish, falling below 10‰ at times), but it also varies with the state of the tide and with respect to depth. Since sea water enters estuaries during high tide, high-salinity water occurs well upstream at this time, only to move seawards again during low tide. Therefore, any given point along an estuary experiences fluctuating salinities through each tidal cycle (Fig. 13). Sea water is more dense than fresh water and thus outflowing fresh water will tend to float as a layer on top of such salt water as is present, with a zone of mixture and of rapid salinity change – the 'halocline' – between the two (Fig. 14). Salinity will then also vary (increase) with depth in the water column, though for this reason not necessarily uniformly. In some estuaries, however, for a variety of reasons the water column is well mixed and in these salinity at any given point is homogeneous from surface to bottom.

Because of their restricted connection to the sea, the salinity of most northwest European lagoons does not vary to such a marked extent as in estuaries either spatially or diurnally, although it may vary seasonally with the evaporation/precipitation ratio, being lower in winter than in summer. Clearly, however, if there is fresh water inflow then salinities will be low in its vicinity, and if sea water input is pulsed, as in those lagoons that receive it only during high spring tides, then, dependent on the relative

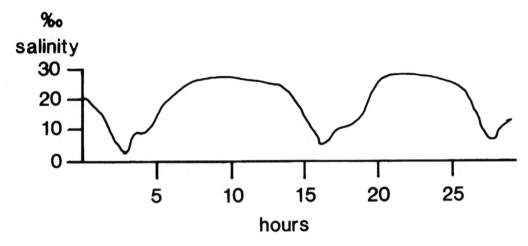

Fig. 13. Tidal variation in salinity at a single point within an estuary (the Pocasset, USA).

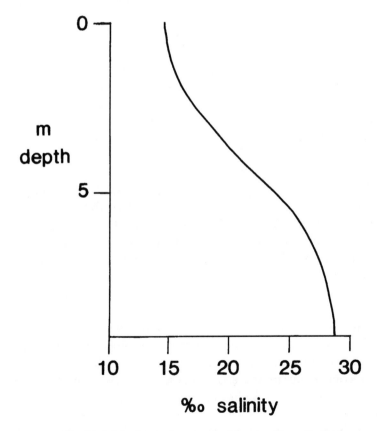

Fig. 14. A halocline in the estuary of the river Tyne, England.

volumes of lagoon and inflow, the salt content of the water will reflect the cycle of input to various extents. But although the salinity of any one lagoon may be relatively constant, the salinity of different lagoons varies enormously – throughout the whole range encompassed by brackish water – as a result of the relative contributions of fresh and salt water. Even adjacent lagoons may possess widely different salinities. Most lagoons are very shallow (see below) and in them there is no variation of salinity with depth. A few are deep, however, and if any of these possess a high level, open connection with the sea, it is likely that the inflowing sea water, being heavy, will descend straight into the depths and there sit beneath the fresher surface water (Dorey *et al.*, 1973). Until churned by strong winds, there may be little tendency of the two water masses to mix, and the one will float on the other and be separated by a marked halocline (Fig. 15). Just as in a deep fresh water lake with a functionally equivalent thermocline, the bottom waters may then periodically be anoxic. Such circumstances, however, are relatively unusual. Although as yet little studied, it is becoming clear that the pH of lagoonal waters is also highly variable, both from lagoon to lagoon and with time in any one system.

Since 1922 (Redeke, 1922) several western European authors (e.g. Aguesse, 1957; den Hartog, 1964, 1974; Amanieu, 1967; Heerebout, 1970; de Jonge, 1974) have endeavoured to erect classifications of brackish waters, particularly landlocked ones, largely on the bases of their salinity on the one hand and of their fauna and flora on the other. Indeed, a whole week-long symposium devoted to this subject was organized by the International Union of Biological Sciences in Venice in 1958. None of the various classifications proposed has been entirely successful, however, not only because of major differences in the faunas of estuaries and largely landlocked lagoons and brackish ponds, but also because brackish waters cover such a diversity of habitat types and within them each species has its own individual distribution pattern in relation to salinity. Nevertheless, it remains true as one might expect that animals of fresh water ancestry are a more significant component of the more dilute systems, and that essentially marine species are more characteristic of high salinity regions and of those with an open connection to the sea (see below). Although the classificatory terminology (e.g. oligohaline, α and β mesohaline, polyhaline and euhaline) frequently appears in the literature, it will not be used here; nevertheless readers may care to note a recent North American analysis of the estuarine salinity gradient (Bulger *et al.*, 1993) that establishes, on biological grounds, the following overlapping zones, in ‰: 0–4, 2–14, 11–18, 16–27 and 24–marine.

An environmental factor that usually varies with salinity is the extent to which the water level fluctuates. For the reasons outlined above, estuaries exhibit dramatic tidal changes in water level and therefore possess extensive intertidal zones. But being less or not at all subject to the tides in the adjacent sea, lagoonal water levels do not fluctuate in the same fashion and most lagoons have no intertidal zone at all. Some small lagoons situated entirely within barrier shingle systems and containing pure sea water as a result of percolation may have water levels that fluctuate with the tides, although usually out of phase because of the time delays inherent in the slow transfer of percolating water (Barnes & Heath, 1980), and those lagoons that do have an open connection

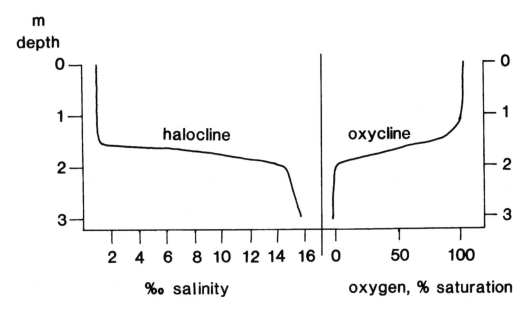

Fig. 15. A lagoonal halocline: that within Swanpool, Cornwall, England. Note the absence of oxygen below the halocline.

with the sea will show tidal fluctuations near their inflow/outflow channels, but these are relatively rare exceptions.

The free-water-mass habitat is exploited, as in many aquatic systems, by plankton, swimming fish and crustaceans, and in some by insects. In the land-locked brackish systems, fish and the crustacean prawns and mysids of necessity spend their entire lives within any given pool, but in those water bodies with an open connection to the sea there is considerable movement between the two. Some species migrate into the sea for the duration of the breeding season, and both adults and juveniles of many more species migrate into brackish waters to feed (see below). Without the facility of at least a hand-operated seine net, it is not possible adequately to sample the adult fish populations, but a pond net of some 1 mm mesh will permit the arthropods and juvenile fish to be examined.

Although salinity and co-varying factors are the diagnostic features of brackish waters, they are perhaps not those that most determine the nature of brackish-water habitats from the viewpoint of many animals. The more characteristic species are markedly eury-haline and can survive anywhere within a range in the order of 4–40‰. Of greater sig-nificance is that estuaries, and even more so lagoons and brackish ponds, are quiet envi-ronments, sheltered from wind action by the surrounding land. And wherever it is quiet in the coastal zone, fine sediments – silts and clays – tend to settle out of suspension in the water column. By Stoke's Law, small particles take a much longer time to fall a given distance through still water than do large ones. Whereas a sand grain of 0.5 mm diameter falls through some 20 cm of water (at 15°C) in 1 s, a silt particle of 0.005 mm diameter falls through only some 0.02 mm in the same time. Fine particles are therefore kept in suspension in wave- and current-agitated sea water and can only settle out in still waters, such as during slack tide in estuaries. Fine particles held in suspension in fresh water also tend to repel each other because of their surface charges; when they are discharged into estuarine waters these surface charges become neutralised and the particles clump together, become larger and hence fall more rapidly. Further, when the particles reach the estuarine bed they may be trapped in mucus-like materials secreted by diatoms. Add to this the stilling and trapping effect of intertidal plants such as those of salt-marshes, and it is no wonder that estuaries are essentially soft-sediment habitats. Estuaries are dominated by soft muds that are often waterlogged and treacherous: readers are advised only to venture onto estuarine muds with extreme caution. The effect of the marine salts in causing fine particles to clump together can be seen in dramatic fashion when heavy rain strikes a muddy salt-marsh. The rain washes away the surface salts; the particles start to repel each other again and disaggregate; and the water flow leaving the marsh is brown with (re)suspended sediment.

In general, lagoons receive a much lesser input of inflowing water, and that arriving via percolation will have been filtered by passage through the isolating barrier. Hence lagoons are often floored by coarser sands and/or shingle (Table 1). Nevertheless, run-off from surrounding marsh areas and, where it occurs, stream inflow will contribute fine particles, and die back and decomposition of fringing and submerged vegetation (see below) may supply copious quantities of fine organic debris to the lagoon bed. Thus, lagoonal sediments usually include a surface layer of fine, highly organic silts, which because they are not tidally exposed to the air and the periodic drying that results, are soft and 'fluffy' in comparison with the more consolidated and sticky sediments of estuaries. Organic matter comprises about 12% of their weight (Bamber et al., 1992).

The finer the particles in a sediment the closer they can pack together and the greater the proportion of the interstitial spaces between the particles that is occupied by non-replaceable water (held to the particle surfaces by capillary forces). Therefore the finer the particles the lower the permeability of the sediment to water: water can pass through a sand filter but not through a layer of silts and clays. Various bacteria can subsist on the organic debris that is deposited with, and becomes buried with, the fine inorganic particles, and in doing so they remove oxygen from the water surrounding them. If they remove oxygen faster than it can be replaced by diffusion through the relatively impermeable sediment, oxygen tensions fall and the conditions are right for the lifestyle of a suite of anoxic bacteria that can respire the oxygen present in the sulphate molecule, and sea water contains abundant sulphate ions (almost 8% by weight of salts). As a result, however, they liberate sulphide ions and hence impermeable muds are black, anoxic, sulphurous and poisonous only a few millimetres below the surface (see Fenchel & Riedl, 1970). Such characterises almost all estuarine sediments and those of muddy lagoons. Even the beds of sandy lagoons are anoxic a few centimetres below the surface since although they are permeable there is no tendency for water to flow through them and utilization of oxygen eventually exceeds diffusional supply.

Several animals, principally crustaceans and gastropod molluscs, roam over the surface of these sediments and only bury themselves shallowly in it for concealment or, in

Table 1. *Percentage gravel/sand/silt content of sediments from various East Anglian lagoons*

	Gravel & cobbles > 2 mm	Coarse sand 2–0.6 mm	Medium sand 0.6–0.2 mm	Fine sand 200–62 µm	Silt < 62 µm
Shingle Street 'lagoon 4', Suffolk	98	0	0	0	1
Holkham Salts Hole, Norfolk	0	31	65	3	1
Dunwich, Suffolk	1	1	84	12	3
Broad Water, Norfolk	0	0	35	46	18
Shingle Street 'lagoon 6', Suffolk	0	1	1	5	93

estuaries, during low tide. Many species on the other hand bury themselves deeply or construct deep burrow systems, including down into the anoxic regions of the sediment (Schäfer, 1972). Those that are merely buried, such as some sea-anemones, worms and insect larvae, usually occur only in the oxygenated surface layers, but life below this zone is made possible for burrow-dwelling worms, for example, by the creation of currents of water through functionally U-shaped burrows. A water current used for respiration and/or feeding will bring with it dissolved oxygen and this oxidises the poisonous sulphides in the burrow wall and converts them back to harmless sulphates. In a spadeful of mud, the deep burrows therefore stand out as brown tubes against the black matrix of the rest of the sediment. The only burrow systems that have surface structures obvious to the naked eye, however, are those of the lugworm *Arenicola marina*. This constructs an L-shaped tube, the vertical limb of which extends down from the surface to join the horizontal limb in which is situated the head and anterior part of the worm's body (Fig. 16). (Muus (1967), Green (1968) and Schäfer (1972) describe the general biology of this and many other brackish-water organisms; see also Fish & Fish (1989)). As the lugworm eats the sand in front of it, so more collapses down to fill the hole, leaving a depression at the surface that marks the site of the worm's head within the sediment. The sand that has passed through the gut is periodically voided at the surface as a faecal string, the worm backing up the vertical shaft

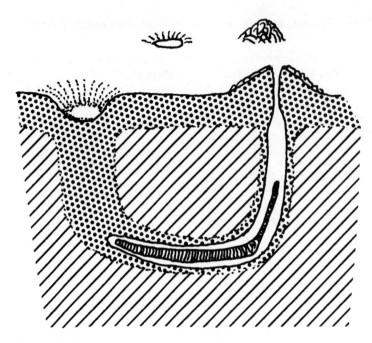

Fig. 16. The lugworm, *Arenicola marina*, in its burrow.

to achieve this. The opening to the burrow soon becomes concealed by coils of this 'string'. In tidal environments, the incoming tide smooths out the faecal mound to greater or lesser extents, but in lagoons where there is little water movement the voided sand can accumulate into volcano-like mounds and the head-end depression can grow into a large conical pit (Fig. 17). Whilst extensive regions of estuarine or lagoonal bed may appear to be featureless expanses of monotonous sand or mud, broken only by surface-dwelling mudsnails and the activities of lugworms, in some spots surface relief is created by the existence of beds of mussels (*Mytilus*) in estuaries and bryozoan reefs in lagoons. Both these may provide a subsidiary habitat for smaller organisms in much the same way that coral reefs do so in the tropics.

Larger particles are not without their biological interest in that they provide a place of attachment for sessile species. Surface shingle, for example, may be encrusted by lagoonal bryozoans or by estuarine barnacles, and pebbles or shells beneath the surface can be used by sea-anemones that extend tentacles over the sediment surface whilst feeding but which can retract into the sediment for safety. Along the barrier margin of lagoons in the southwest of the region, the interstices of the shingle itself may provide a specialist habitat for a group of small and in several cases rare molluscs – *Truncatella subcylindrica*, *Paludinella littorina*, and *Caecum armoricum* (see Little *et al.*, 1989) – and the ribbon-worm *Prosorhochmus claparedii* occurs beneath the pebbles in the same areas. Somewhat more widely distributed in the same microhabitat are the snail *Onoba aculeus* and the sipunculan *Nephasoma minuta*.

For sampling the bottom-dwelling animals, a scoop of known surface area and a sieve are all that is needed. The author uses scoops of 0.01 m² area and 2 cm or 5 cm depth, and 500 μm and 710 μm mesh sieves dependent on the coarseness of the sediment. Sediments containing abundant shingle may first have to have the latter removed with, e.g., a 2 mm mesh sieve. Sieving is best achieved by gentle puddling in a source of water: this is no problem in non-tidal habitats, but it can be so in estuaries at low tide. The contents of the sieve should be gently washed into a white tray for observation. A scoop of 5 cm depth will clearly not sample the deeper burrowing worms and bivalve molluscs. The presence of *Arenicola* will be obvious from its characteristic surface marks (see above), but although some of the other deep-burrowing species do produce equally diagnostic if less obvious traces, for example the bivalve *Scrobicularia* (Fig. 17), their presence can only really be verified by use of a spade (or in respect of the molluscs from the presence of empty shells nearby – although this evidence should be treated with caution in estuaries since shells are long lasting and can be transported considerable distances by currents). Studies of the effect of bait-digging have shown that digging causes serious long-term habitat disturbances and it should only be carried out in moderation (see Davidson *et al.*, 1991) and never in small lagoons and ponds. All species dig deeper burrows as they grow, and often during winter, and so it may be possible to determine the presence of a given animal by finding young individuals in the surface sediments. Juvenile *Nereis diversicolor* are often abundant in the uppermost 2 cm, although adult burrows extend down 40 cm.

Another feature common to northwest European brackish waters is their shallow depth. Excepting the longitudinal channel that may be dredged to provide access for shipping, estuaries are clearly shallow: most of their area is exposed by low tides. The majority of lagoons are even shallower: few have depths greater than 1 m. There are of course exceptions. Those that were originally gravel pits within shingle formations are likely to be deep (>3 m), and those that were once small estuaries or areas of salt-marsh may have deep areas in the region of former major creeks and channels. The same applies to all the various types of brackish ponds; naturally formed ones are usually shallow, but those in depressions excavated by man can be expected to have maximum depths of over 3 m. There are at least two reasons for the prevailing shallowness.

One is that their basins have been filling with fine sediments, organic matter, and coarser material washed over from the enclosing barriers by storms. Cores taken through the beds of brackish systems indicate considerable thicknesses of accumulated sediments, interspersed in the case of lagoons by wash-over fans of shingle (Fig. 18).

The other is that percolation-supplied lagoons and pools are limited in height (i) by the height of the water table within the barrier (which will be well below high tide level in the adjacent sea), (ii) by the level of the outlet system (which is usually via a drainage ditch and therefore will be low in the interests of land drainage), and (iii) by the topography of the flooded land, which is most likely to be former grazing marsh that was (a) relatively flat and (b) already at a high level – high water neap tide level or higher – so that no great depth of water can accumulate. All of these conspire to keep percolation-fed pools, including the scrapes in bird reserves, below some 0.5 m in depth. As a rule of thumb, the larger the system the deeper it is likely to be (although beware small former gravel pits!), at least in its centre, and pools of regular geometrical shape bounded by earth walls can be expected to be too deep to wade across.

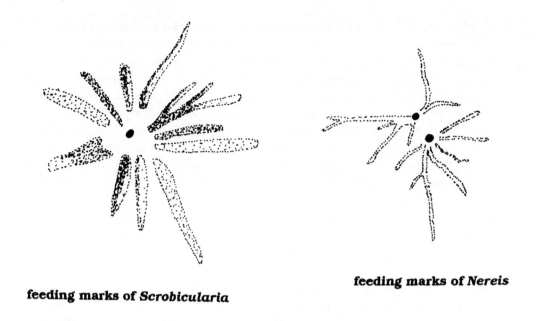

feeding marks of Scrobicularia

feeding marks of Nereis

cast and depression of Arenicola

Fig. 17. The surface marks indicating the burrows of *Scrobicularia, Nereis* and *Arenicola* within the sediment.

Brackish-water habitats are therefore shallow and floored by soft sediments. This in turn encourages the growth of fringing stands of vegetation: reedbeds in the case of pond-like systems and salt-marsh along estuaries. Indeed, such is the shallowness of most lagoons that reedbed can, if not checked, extend right across the water mass, eventually to obliterate all free water and to hasten a natural succession from marine pond to woodland, via reedbed and alder carr. The shallowness of bodies of brackish water also means that light may penetrate right down to their beds, permitting the growth of algae and rooted but submerged plants. In estuaries, these include eel-grass (*Zostera*), wracks (*Fucus*) and thin-walled green seaweeds (*Ulva* and *Enteromorpha*),

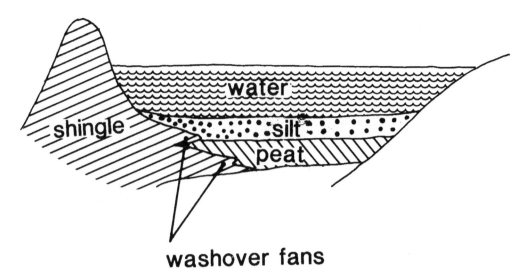

washover fans

Fig. 18. A diagrammatic section through the shingle barrier enclosing a lagoon, showing washover fans.

whereas in lagoons and brackish ponds they may be joined or replaced by pondweeds (*Potamogeton pectinatus* and *P. filiformis*), holly-leaved naiad (*Najas marina*) and horned pondweed (*Zannichellia palustris*) in the less saline systems, and by tasselweeds (*Ruppia maritima* and *R. cirrhosa*), charophytes (particularly *Lamprothamnium papulosum*) and tough, wiry, filamentous algae like *Chaetomorpha* in the more saline ones (Fig. 19).

In lagoons, at least, the submerged and fringing vegetation is extensively colonized by animals, including, somewhat surprisingly, juvenile cockles: most lagoonal species are associated with the submerged tasselweeds, etc., at some stage of their life history (Verhoeven, 1980). Adult insects (waterboatmen and beetles) and prawns swim through the spaces between the individual leaves. Gastropods and the crustacean isopods and amphipods crawl over the plants. Some insects and amphipods dwell in tubes cemented to the stems and leaves, whilst hydroids and young cockles attach to them, and bryozoans encrust the larger stems, particularly those of reeds. Some sea-anemones and insect larvae may simply entwine their bodies in amongst the algal filaments. In estuaries, the equivalent eelgrass and seaweeds are alternately exposed and submerged, and perhaps hence are less abundantly colonized, but they do provide shelter for amphipods and others during low tide; the fringing salt-marshes, however, may be a rich habitat for insects, molluscs and crustaceans. We know relatively little of the position in life and the life styles of the associated animals, particularly in respect of the submerged vegetation. Such habitats have mainly been investigated, insofar as they have been investigated at all, by ecologists concerned with population dynamics and the like, and the sampling techniques used have provided little or no information on the precise spatial arrangements of the individuals. Here is a field in which the naturalist can make many important observations relevant to the biology of the animals concerned.

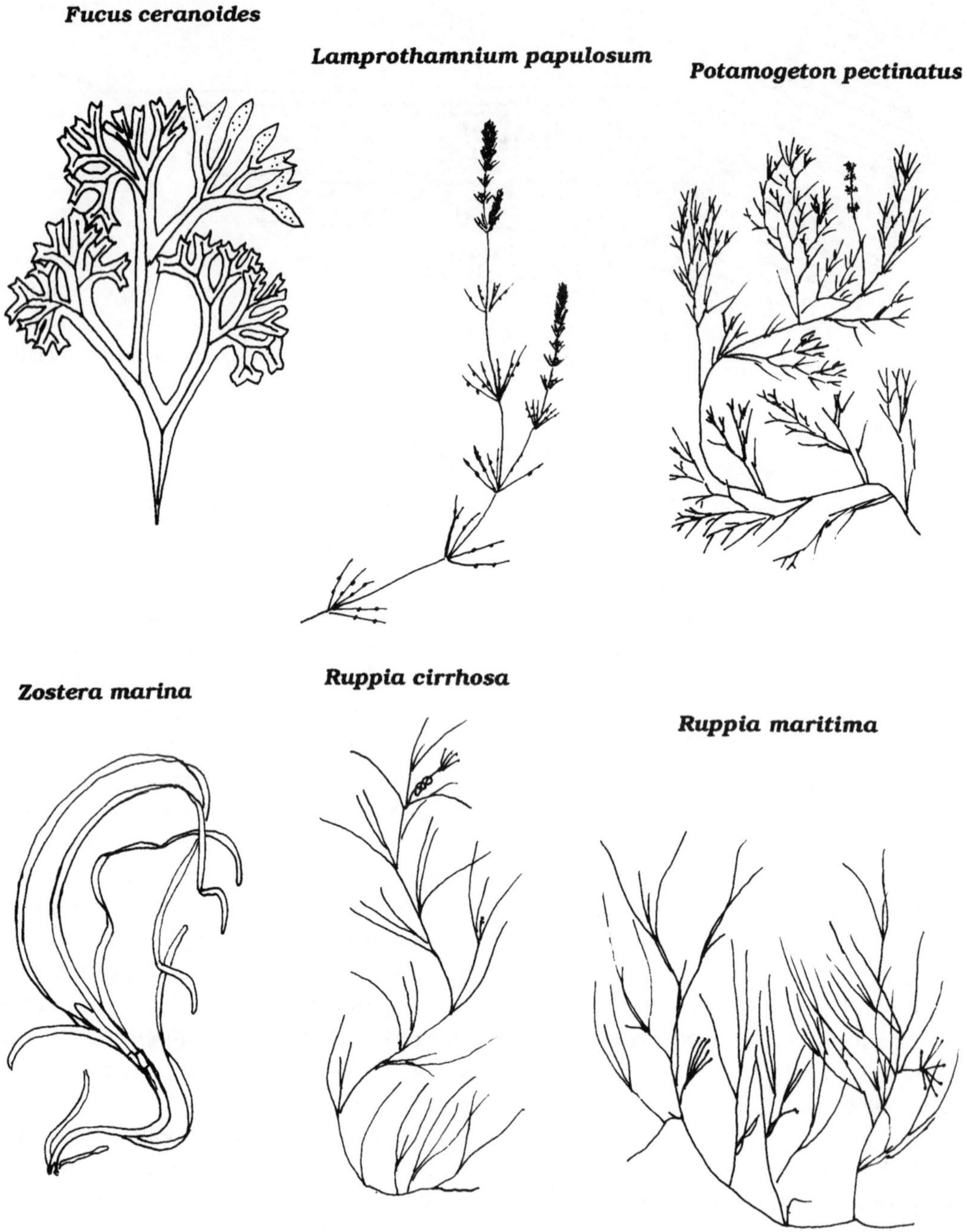

Fucus ceranoides

Lamprothamnium papulosum

Potamogeton pectinatus

Zostera marina

Ruppia cirrhosa

Ruppia maritima

Fig. 19. Characteristic submerged macrophytes of brackish waters.

The presence of animals encrusting the stems of reeds can be determined by direct inspection (including of wooden posts and other structures sunk into the sediment), and the same applies to those building tubes on the submerged vegetation. The free water mass within stands of submerged and emergent plants poses greater problems, however, and is best tackled – at the start at least – by attaching a flour sieve to a long pole and sweeping it through and between the plants. A real or makeshift shepherd's crook can be used to bring vegetation nearer to the observer, again taking great care not to damage rooted species. Access to a boat would clearly ease this problem, and make direct observation possible.

The above account has concentrated on the habitats available to the larger animals, but of course many animals smaller than 0.5 mm are also present in brackish waters. They have been omitted only because they require expensive apparatus and specialist knowledge for their collection and identification. Nematodes, small flatworms, copepod (especially the harpacticoids) and ostracod crustaceans are abundant in all but the most anoxic habitats in brackish waters, and ostracods in particular do span the somewhat arbitrary 0.5 mm division separating the larger from the smaller animals. In respect of this group, the reader is referred to Athersuch *et al.* (1989) for information on their biology.

It is in many ways convenient to treat these water, sediment and vegetation habitat types as if they are separate – not least because they require different techniques for their investigation – but they are not really so. Several species occur in all of them, in some cases readily moving from one to another, and juvenile and adult lives may be spent in different parts of any single system, including of course those bottom-dwelling animals that have planktonic larvae. The sea-anemone *Nematostella* can occur both in the sediment and amongst the submerged vegetation, for example, and the bryozoan *Conopeum seurati* is as likely to encrust surface shingle as reed stems, or indeed pieces of wood floating in the water. The goby *Pomatoschistus microps* swims in the open water and in amongst the *Ruppia* and reeds, and it feeds at the sediment surface, where it also lays its eggs (within empty cockle shells).

Nature of brackish-water faunas: estuaries

The faunas of the brackish-water stretches of estuaries are mainly impoverished versions of that marine fauna that inhabits shallow areas of soft sediment. Specifically, they comprise those essentially marine species that can withstand low and fluctuating salinities. Some clue as to why they might inhabit this seemingly hostile habitat is provided by their differential distribution in the faunistically rich southwest of the region and in the relatively species-poor southern North Sea basin. In the drowned river valleys of the southwest, a diverse array of marine species occupies the adjacent sea and the high salinity zones of the estuaries, confining the brackish-water fauna to the inhospitable regions of soft muds and widely varying salinity. In contrast, in the North

Sea from which the species-rich southwestern fauna is absent, the supposedly brackish-water species occur not only throughout the estuaries but also in the coastal sea itself, wherever sheltered conditions prevail (Fig. 20). It would appear that in the southwest, the 'brackish species' can only colonize those areas that lie outside the ranges of the host of superior marine competitors also present.

Competition, however, would not seem to be the process limiting the upstream penetration of a significant number of the estuarine species. Neither do their upstream distributions consistently relate to any obvious feature of the environment, such as salinity, as one might expect. Yet the salinity of the water must have some effect because very few of these essentially marine species can withstand fresh water. It appears most likely that any attempt to correlate the observed ranges of most of them with specific environmental variables is futile, in that their distributions are dynamic and ever changing, as is the estuarine environment itself. At times of maximum fresh water discharge, resulting from severe storms, etc., the upper reaches of estuaries virtually become rivers, and widespread mortality occurs amongst the local brackish-water fauna. When the estuary returns to 'normal', the ousted species slowly recover the lost ground by individual dispersal or via reproduction. How far upstream they happen to have got at any particular point in time will depend on when they were last affected by a period of greatly lowered salinity, on their rates of potential dispersal, and on the timing of their breeding. They may well not recover their former ranges before being locally wiped out again. Only in systems with a constant salinity regime, like the Baltic Sea, are their upstream distributions likely to be stable and not in a state of continuous forced retreat and expansion.

This state of continual local extinction and recolonization has also been taking place on a much grander scale over the last 2 500 000 years. When northwest Europe was either dry land or glacial ice caps, the rivers of southern and southeastern England, northern France, the Low Countries and northern Germany, together with melt water from the glaciers themselves, must all have discharged into a huge river passing along what is now the middle of the English Channel to empty into the Atlantic in the Western Approaches (between Brittany and the Scilly Isles). As sea level rose again, a large estuary must have moved up the Channel as far as the south-western North Sea. Wolff (1972; 1973) has argued that the large brackish area that must have existed in the Western Approaches at the mouth of this enormous river prior to 20 000 years B.P. was the place of origin of the local brackish-water fauna. It may well have been a refuge for the local estuarine fauna, but it seems unlikely that it would have accommodated non-tidal, lagoonal species. Lagoons are destroyed (emptied) by falling sea levels and only created by rising ones, so that any species restricted to non-tidal habitats would have already become locally extinct during a glacial phase, and neither would the Western Approaches estuary, being tidal, have been an appropriate habitat for them (see below).

One consequence of this series of repeated invasions of the North Sea basin by estuarine faunas, followed by repeated partial or total southwesterly retreat, is that during each period of colonization an individual marine species might have given rise to a new local brackish-water form, and if say five recolonizations took place, species A

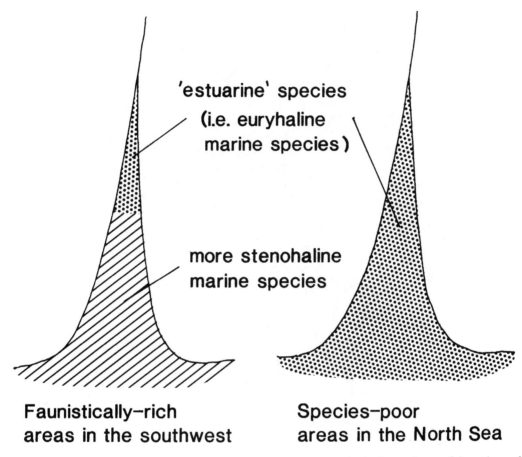

'estuarine' species
(i.e. euryhaline
marine species)

more stenohaline
marine species

**Faunistically–rich
areas in the southwest**

**Species–poor
areas in the North Sea**

Fig. 20. The contrasting distributions of typically estuarine species in the southwest of the region and in the North Sea.

might have provided the ancestry of species B, C, D, E and F, one on each separate occasion. This may have happened in the amphipod *Gammarus*, in which *G. duebeni*, *G. zaddachi*, *G. salinus*, *G. oceanicus*, *G. insensibilis* and others are all closely related to, and are thought to have derived from, the relatively marine *G. locusta*. Speciation, however, is not yet complete. At the start of the breeding season the larger male *Gammarus* take hold of the smaller females and swim with them in 'precopula' pairs (Fig. 21). Insemination takes place when the female moults, although the male still maintains precopula for a time thereafter, thereby denying access to competing males. But, in the laboratory at least, males of these species frequently and mistakenly take possession of females of the wrong member of the species flock (Kolding, 1986). Fertilization occurs but the hybrid embryos all die at some stage of their development (between the 256 cell stage and the first juvenile moult), dependent on the precise interspecific pairing. This means that such mistakes are very costly, the perpertrators leaving no offspring. In nature therefore there would be strong selection pressure against such inappropriate attempted matings, and what seems to occur to prevent this is that the various species

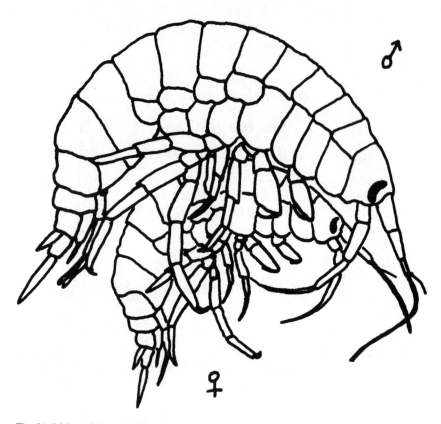

Fig. 21. Male and female *Gammarus* in precopula.

separate themselves along some convenient environmental gradient. In addition, adjacent species change their breeding season so as to reproduce at different times of the year. In most areas the spacing-out gradient adopted is that of salinity (even though there are no physiological or ecological necessities so to do), so that each species occupies only a very limited segment of any common salinity gradient (Fig. 22); but where all species occur at the same salinity, as in the Baltic, they divide up the depth and microhabitat gradients in a comparable fashion instead.

Nature of brackish-water faunas: lagoons

Several members of this estuarine fauna also occur in lagoons and equivalent brackish ponds (e.g. the ragworm *Nereis diversicolor*, the lugworm *Arenicola marina*, the amphipods *Corophium volutator* and *C. arenarium*, the winkle *Littorina saxatilis*, and the Baltic tellin *Macoma balthica*), but there they are joined by two other faunal components. The first of

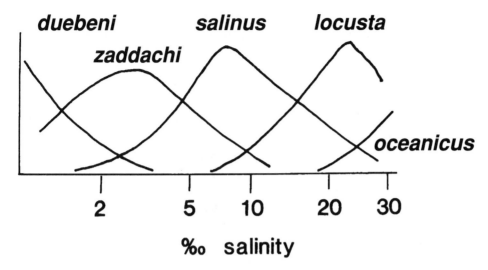

Fig. 22. The distribution of *Gammarus* species along the salinity gradient of the Limfjord, Denmark.

these is terrestrial or fresh water in ancestry. Numerous insects (particularly dipteran larvae and all life-history stages of beetles and corixid and notonectid waterboatmen) and several oligochaete worms are especially characteristic of coastal bodies of water, and in the most dilute of coastal brackish waters they may be joined by animal species that are otherwise typically pond dwelling, for example pond snails (*Lymnaea*) and the hog louse (*Asellus*). The second component does not occur in fresh waters, estuaries or the sea, but is restricted to lagoonal and similar habitats. It is clearly related to the estuarine/marine fauna, however, in that the two faunas contain a number of species-pairs, one species being typical of each environment. The marine edible cockle, *Cerastoderma edule*, is replaced in lagoons by the closely related *C. glaucum*, for example, and the small mudsnail *Hydrobia ulvae* by either or both of *H. ventrosa* and *H. neglecta*. Other species known only from lagoons, at least in northwest Europe, include the hydroid *Clavopsella navis*, the sea-anemone *Edwardsia ivelli* (possibly now extinct, having apparently disappeared from its only known locality in the world, Widewater lagoon in Sussex, U.K.), the polychaete worms *Armandia cirrhosa* and *Alkmaria romijni*, the amphipod *Gammarus insensibilis*, the winkle *Littorina saxatilis lagunae*, and the bryozoan *Victorella pavida* .

Since these specialist lagoonal species do not occur in the coastal seas of northwest Europe, a question that begs an answer is: How do they disperse from one lagoon to another? The answer must clearly be: only with great difficulty. This may be evidenced by the extreme species poverty of some lagoons: numbers of suitable niches appear to be present but they remain unfilled. Their faunas are 'unsaturated'. One potential dispersal mechanism that has been popular ever since Charles Darwin is birds' feet. Waterfowl and waders may occasionally get a piece of alga or waterweed entangled around a foot, and that piece of weed may have associated with it the eggs or juveniles of aquatic animals. Thus, when the birds fly to another pool, the aquatic animal hitches an involuntary lift. If that is so, then lagoons that are much frequented by birds would be predicted to have a common pool of transportable and transported species, and have faunal differences from bird-less lagoons. This was not borne out by the investigations of this author (Barnes, 1988): the faunas of bird lagoons were no more similar to each other than were those of bird-less lagoons, or than the similarity between birdy and bird-less ones. An alternative hypothesis revolves around the periodic invasions of the land by sea water mentioned above. Algae and waterweed debris float, and so coastal floods will tend to disperse this vegetation, and its associated animals, from area to area. If this is the major means of dispersal then recent heightening of coastal defenses will have rendered many land-locked brackish-water communities relict. Local extinctions can still occur but now little or no balancing immigration is possible. Without compensatory measures we may therefore witness the decline to extinction of lagoonal species such as *Hydrobia neglecta* and *Cerastoderma glaucum* in northwest Europe.

If the specialist lagoonal species do not occur in the estuarine or marine environments yet are clearly related to estuarine/marine forms, a second question arises: What was the origin (or origins) of this fauna? This is still contentious, and any answer must take into consideration the following observations. The fauna is not in fact a brackish-water one (Guélorget & Perthuisot, 1983; Bamber *et al.*, 1992; Barnes, 1994a), although it is clearly capable of living under such conditions. And whilst it does not occur in the northwest European seas, it is found in the Mediterranean basin and elsewhere (see Guélorget & Perthuisot, 1992, for example); 'elsewhere' includes, for some species, the Black and Caspian Seas as well as the Arctic and the American shores of the Atlantic. The common factor in the above appears to be that these species are particularly associated with tideless environments, whether brackish or fully marine, and they are restricted to lagoonal habitats in northwest Europe simply because lagoons are the only tideless maritime habitats available there.

To understand this association with tideless conditions we must again return to a historical perspective. Some 100 million years ago, a large ocean ('Tethys') extended from what is now southeast Asia through the present basins of the Caspian, Black and Mediterranean Seas to open into the Atlantic, which at that time was closed to the north since western Europe was then still joined to north America. The drifting movements of Africa and India over the next tens of millions of years sealed off the Tethys, which degenerated into a series of land-locked, marine to brackish seas in the general area now occupied by part of the Mediterranean, the Black and the Caspian Seas; the Mediterranean was sealed off to the east 20 million years B.P. It therefore seems possible that this fauna evolved in that largely land-locked and therefore tideless sea(s). How it reached northwest Europe is more problematic. It might seem obvious that it did so via the Straits of Gibralter, but the Mediterranean has been isolated from the Atlantic at various times during the last 15 million years, and between some 5.5 and 4.8 million years ago it twice became completely enclosed and dry (its water having evaporated thereby leaving the large deposits of salt that underlie the present sea). Any aquatic organisms living in it as it began to dry up would almost certainly have died out. (To account for the quantity of deposited salts, it is clear that water must have continued to flow from the Atlantic into the Mediterranean, although not vice versa. Even today

Atlantic water flows into the Mediterranean to balance the rate of net evaporation, which is currently running at over 3 000 km³ of water a year.) Any origin in the Mediterranean region must therefore have been either before 15 million years B.P. (much the more likely) or after 5 million years B.P. Another possibility could be a route from the present Caspian Sea area into the Arctic Ocean at the White and Kara Seas, since a series of inland brackish seas may have linked the Arctic and Tethys Oceans between about 150 and 50 million years B.P. All these millions of years B.P. may seem to be much too long ago to qualify for the origin of a fauna alive today, but the available evidence indicates that several brackish-water species are very old indeed. The characteristic gastropod genus *Hydrobia* dates back 170 million years, the Caspian fauna could be 300 million years old, and it has been claimed that the characteristic bivalve genus *Cerastoderma* originated in the Ordovician, some 450 million years ago.

In any event, after the fauna colonized tideless maritime environments in northwest Europe (and onwards to northern America), it would have been eliminated from this area during the glacial phases of the Pleistocene, when the whole of the region was either land or under ice. Probably, it was able to find a refuge in the Mediterranean (which was open to the Atlantic throughout the Pleistocene) each time the ice advanced and to move back again into northern Europe via the lagoon systems that would have been created as sea level advanced and the ice retreated. Why these species should be unable to endure life in tidal seas is a mystery however.

The diets of brackish-water animals

Two features held in common by the estuarine and lagoonal faunas, regardless of their origins, are that they comprise a very limited array of feeding types and that many of the species are extremely generalist in their diets. This can be exemplified by the rag-worm, *Nereis diversicolor*, which is a common component of both faunas. The ragworm (a) will catch and consume other animals smaller than itself, (b) will eat pieces of green seaweeds, (c) will scavenge, even pulling small dead fish into its burrow, (d) will consume the surface layers of the sediment, and (e) can secrete a cotton-wool-like filter of mucus in the upper part of its burrow and then by undulating its body cause water to flow through the filter, thereby trapping plankton and other suspended particles, and then ingest the whole mucus plus particles system. Some of these feeding modes are likely to provide a richer source of food, for every unit of energy expended in obtaining it, than are others, and accordingly the ragworm has a heirarchy of preferred modes. Granted the opportunity so to do, it will scavenge nearby animal carcasses preferentially; next comes catching small shell-less animals, then, in descending order, browsing the sediment surface, using its mucus filter, eating algae such as *Ulva* and *Enteromorpha* , and finally hunting live shelled gastropods. This order of decreasing dietary interest exactly corresponds with that of decreasing value of the prey (Pashley, 1985 – cited in Barnes *et al.*, 1988).

In microcosm, the ragworm illustrates all the feeding modes of brackish-water animals: carnivory/scavenging; filter-feeding on particles suspended in the water; and deposit-feeding on materials on and in the surface layers of the sediment. Filter-feeding is less prevalent in brackish environments than it is on sandy or rocky marine shores, perhaps because the especially high silt content of estuarine waters would tend to clog any filter and render food collection and sorting energetically expensive. Nevertheless, bivalve molluscs such as the cockles, using their gill as the filter, crustaceans with setae on their mouthparts, and various animals with ciliated tentacles, such as the bryozoans and a number of polychaete worms, are relatively specialist suspension-feeders, even if they may rely to a considerable extent on water movements resuspending already deposited materials for their food particles. Oysters (usually introduced species of *Crassostrea*) have been laid to 'fatten' in a number of estuarine mouth areas, and they have been recorded in waters turbid with silt as ceasing to endeavour to sort the potential food particles from the background silt, but simply swallowing the lot. Rather more commonly, bivalves with gill filters or worms with ciliated tentacles may either suspension feed or deposit feed as circumstances dictate. Tentacles may be held up into the water from the safety of a burrow or tube in the sediment, or may equally be extended laterally to roam across and through the surface sediments. In either case suitable food particles can be collected and conveyed back to the mouth via cilia. Unlike the cockles and mussels, many brackish-water bivalves possess two long siphons, the more elongate of which – the inhalent siphon – is used to draw a water current (containing the potential food particles) into the shell cavity, within which it can be filtered by the gill and from which it can be expelled through the somewhat shorter exhalent siphon. Similar to the operation of the tentacles described above, the inhalent siphon can just as easily suck up already deposited particles, like a vacuum cleaner, as draw in a current from the overlying water mass. Exactly the same situation applies to crustaceans with setose filter systems, so that amphipods like *Corophium*, bivalves such as *Scrobicularia*, *Abra* and *Macoma*, and various worms, including the spionids and *Lanice*, can all both suspension- and deposit-feed.

The plankton is an obvious food source for suspension feeders, although what nutriment the deposit feeders obtain from the sediment is less evident. Such animals are usually labelled 'detritus feeders', but this is only begging the question since detritus covers a multitude of possible foods differing greatly in their digestibility. A given flake of detritus is most likely to be a piece of vegetable debris originating from the fringing or submerged plants or algae, or in estuaries brought in by river or high-tide water from more remote sources. After it has been in the water column or on the bottom for any length of time, this flake will also have in and around it bacteria, microscopic consumers of bacteria such as ciliates and flagellates, and other minute, but not quite so microscopic, nematodes, flatworms and copepods. Further, if the flake (or sediment particle or even a piece of living vegetation) is in the light, it will be used as a substratum by living photosynthetic organisms such as diatoms. Thus, a detritus feeder ingesting such a flake or browsing over the submerged plants will take in all these different components: which can it digest? For most brackish-water detritus feeders the most likely answer is that they are dependent mainly on the living diatoms and equivalent photosynthetic flagellates associated with the particles (Barnes & Hughes, 1988). The original piece of vegetable debris (or living stem of *Phragmites* or *Ruppia* or whatever) is likely to be indigestible. The debris will have long since leached and decayed down to cellulose, which the digestive systems of animals cannot tackle, whilst the associated bacteria will usually be too diffuse and at too low a density to satisfy the needs of animal consumers. The diatoms and flagellates, however, are not only digestible, but do occur in sufficient densities to provide an adequate diet. That is not to say that bacteria, ciliates and minute animals are not also digested, in some cases they certainly are, merely that they are not of over-riding nutritional significance.

Most detritus feeders, other than the tentaculate and gill-filtering ones mentioned above, move over or through the sediments or browse the area around the mouth of their burrow and simply swallow all or part of this detrital aggregate. Examples include the lugworm and morphologically equivalent polychaete worms, many of the amphipod, isopod and insectan arthropods, and a large number of the gastropod species. Here again, the dividing line between detritus feeding and carnivory/scavenging is very blurred, being only a matter of size of particle ingested. The prawn *Palaemonetes varians*, for example, swims over the sediment surface collecting food items with its small pincers. The food items concerned may mostly be diatomaceous or detrital (Escaravage & Castel, 1990), but the prawn will also prey on the small burrowing anemone *Nematostella* when this is available (see Posey & Hines, 1991). Specialist invertebrate predators are rather uncommon, although the nemertine worms (e.g. *Lineus*), several beetles, and most of the sea slugs come into this category. Some of the sea slugs are particularly specialist in that the range of prey which they hunt is very restricted. This may account for their rarity in brackish waters. Even characteristically brackish-water species such as *Tenellia adspersa* are thinly scattered both between and within sites, because their hydroid prey *Cordylophora*, *Protohydra* and *Gonothyraea* are likewise far from abundant. The small *Doridella batava* that used to be restricted to the Zuiderzee where it also fed on *Cordylophora* is "almost certainly now extinct throughout its original range" (IUCN, 1983) and probably elsewhere. Even though specialist predators are uncommon, several generalist species are carnivorous when the opportunity presents itself, as noted above with respect to the ragworm and prawn, and we will see below that the impact of some of these unspecialized generalist consumers on the larvae and juveniles of the remainder of the fauna can be huge.

It is the vertebrates that are the top consumers in brackish waters. Indeed, the birds are not only the reason why most people visit estuaries and lagoons, but they are why most protected lagoons and brackish ponds are within nature reserves. Waders and to a lesser extent wildfowl are well known to winter in these areas, where they feed on all the various animals mentioned above, probing the mud with their beaks (curlew, godwit, oystercatcher and redshank), picking over its surface (plovers, etc.) or scything through it when exposed (shelduck) or when covered by a few centimetres of water (avocet). All prey species are taken from the larger worms (by godwit) and bivalves (by oystercatcher) to the small *Corophium* (by redshank and avocet) and *Hydrobia* (by shelduck and the smaller waders). Some small predatory fish (sticklebacks and the goby *Pomatoschistus microps*) permanently inhabit lagoons, but the major impact by fish on estuaries and on those lagoons with open connections to the sea is in their use as nurs-ery grounds, particularly during the summer months. Herring-like fish and several flat-fish have a life cycle in which the juveniles inhabit and feed in brackish waters, and adult grey mullet, flounder and sea bass feed there too. Young flatfish feed unselec-tively on the small and juvenile animals inhabiting the surface sediments, whilst adult grey mullet (one of the rare vertebrate 'detritus'-feeders) and flounder also take larger mouthfuls of the mud. Growth rates fueled by invertebrate productivity can be very large (Fig. 23). The overall pattern of the food web is shown in Fig. 24; it is character-ized by the limited number of comprising species but the great complexity of the inter-relationships between them (Barnes, 1984; 1994b).

What limits the populations of brackish-water animals?

It might seem obvious that the waders and other shorebirds that characterize estuarine areas have a significant impact on their prey, and indeed there is no question that indi-vidual birds feeding on mudflats can consume prodigious numbers of invertebrates each day. A medium-sized wader such as a redshank, weighing some 0.13 kg, can eat in the order of 40 000 *Corophium* per day (Goss-Custard, 1969). A population of 10 000 redshank (as occurs in the Dutch Waddenzee in August, for example) therefore could eat about 400 million of this amphipod a day, equivalent to 15×10^{10} a year. Similarly, the oystercatchers of the same area could eat 45 million cockles each year, and the knot some 13×10^9 Baltic tellins (*Macoma balthica*).

The effect that this enormous consumption has upon the prey populations, how-ever, is almost certainly negligible in spite of the sometimes significant proportions of adult prey individuals taken (Raffaelli & Milne, 1987). For a start, the number of inver-tebrates available is itself colossal. Although the Dutch section of the Waddenzee sup-ports an average of half a million intertidally feeding birds, it also contains in excess of 75×10^{10} *Corophium*; 44×10^9 cockles; and 15×10^{10} *Macoma* just to consider the adults (Wolff, 1983). Densities of *Hydrobia* can exceed 300 000 individuals/m^2 (Smidt, 1951). But there are two important additional reasons why the birds exert an insignificant effect. First, shorebirds feed on relatively large individuals of their preferred prey, and large individuals are usually an insignificant proportion of the total prey population. Thus, in the Gironde Estuary in France, for example, there has already been a mortal-ity of more than 99% before the mudsnail *Hydrobia* ever achieve a size that would bring them within the sphere of interest of the numerous bird species that consume this animal (Bachelet & Yacine-Kassab, 1987). Only 13% of the Gironde mudsnails survive for more than 3 months after metamorphosis, and less than 1% survive for 9 months. Using the largest known percentage mortalities inflicted by birds on mudsnails more than one year old, bird predation could account for less than 0.1% of that to

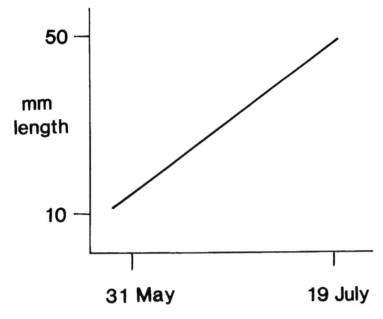

Fig. 23. Growth rate of juvenile flounder, *Platichthys flesus*, over a period of 50 days when in Danish brackish waters.

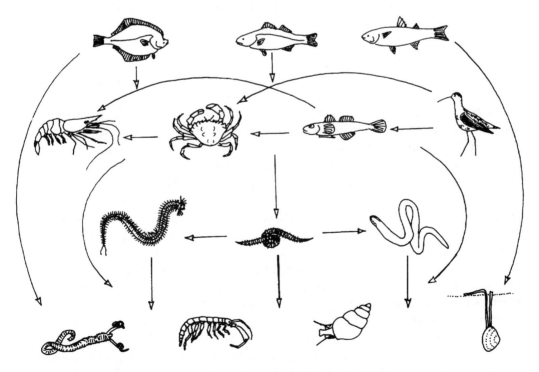

Fig. 24. Highly schematic food-web relationships within the brackish-water macrofauna; the worms, crustaceans and molluscs at the base of the web feed on benthic diatoms, detritus, etc.

which the post-metamorphic snails are subject, and mortality whilst they are larvae is also enormous.

The second reason why bird predation is insignificant concerns the nature of the consumers. Birds are much larger than their prey (a redshank weighs some 100 000 times as much as each of its individual food items), and they are warm blooded and must therefore burn up most of their food just to maintain body temperature. Their food requirements are therefore very large. If the redshank in the example above eats 40 000 *Corophium* per day on average in winter, and if its prey are only available for some five daylight hours a day because of the tidal cycle, redshank will each have to obtain one *Corophium* during each half a second of this period in order to achieve their necessary daily intake. And this is of an animal hidden from their sight, up to 4 cm below the sediment surface!

Not surprisingly then, shorebirds are highly selective both of where they feed and of prey size: they forage as optimally as possible (Piersma, 1987; Zwarts & Esselink, 1989; Zwarts *et al.*, 1992). This means that the effect of the density of their prey on the birds is much greater than the effect of the birds on their invertebrate prey. Feeding must be concentrated in areas of greatest adult prey abundance and its effects are necessarily patchy; areas of even moderate prey numbers may have to be ignored if they cannot supply sufficient acceptable items per unit time. The smaller size ranges of the prey are simply too small to be taken and, in any event, would not be energetically worth the effort (Fig. 25). If shorebirds cannot have a serious impact on their intertidal prey populations in estuaries, it is inconceivable that they could do so on the less accessible invertebrate populations that characterize the permanently water-filled lagoons and brackish ponds. The relationship between bird feeding and invertebrate abundance can be verified, or in some species established, on relatively sandy areas of estuaries (for ease of access) using surface feeding bird species (for ease of censusing the invertebrates). Areas where birds spend a disproportionate amount of time feeding can be established through binoculars, and then the densities and/or size structure of the potential invertebrate prey there can be compared with those in 'neglected' areas.

What effect might the fish have? At least in estuaries, it appears that adult fish have an impact roughly equal to that of the birds (Wolff & Wolf, 1977; Baird & Milne, 1981), and like them they concentrate on large prey items. Perhaps of more importance than the ability of birds and adult fish to cause significant mortality is their cropping of any parts of prey individuals that are located near the sediment surface; for example the tail ends of *Arenicola* and *Heteromastus* as these worms back up their burrows to defaecate, and the filter-feeding tentacular crowns of fanworms and the siphons of bivalve molluscs (Vlas, 1979). The cropped individuals can regenerate the lost portions, but in order to do so they must divert resources away from growth and reproduction, which can have knock-on effects on recruitment.

Several studies have used a caging technique to investigate the effects of surface-dwelling predators. In essence, areas of sediment are protected beneath frames covered by netting of known mesh (large mesh to exclude only large species and small mesh to exclude all but the smallest), and densities of prey species within the cages are then compared, after various time intervals, with those in surrounding regions of unpro-

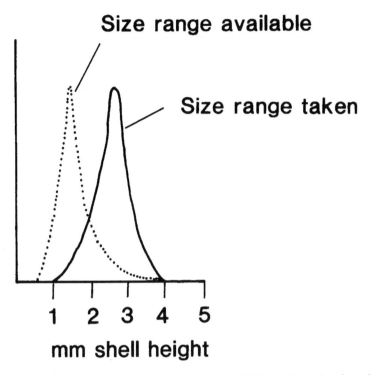

Fig. 25. The size range of prey (the mudsnail *Hydrobia ulvae*) taken by a bird predator (the knot *Calidris canutus*), in the Dutch Waddenzee, in relation to the size range of *Hydrobia* available to it there.

tected substratum. These field experiments have shown that small fish the size of gobies and the juvenile stages of larger fish species (together with small crabs, shrimps and prawns) can have a dramatic effect on the density of sediment-dwelling animals, and even on the numbers of such species present in an area (Reise, 1985) (see Table 2). That table also shows that the adults of deep-burrowing species, such as the worm *Heteromastus*, are least or completely unaffected, as might be expected. In the German Waddenzee, the effect of these small predators is to reduce the numbers of cockles from densities of up to 180 000 individuals/m² in the early summer when the spat settle to almost zero at the end of that summer in eight years out of ten. (It is relevant to later discussion to note that removal of adult cockles has no positive effect on juvenile survival.) These 'epibenthic' predators, as they are called, walk or swim over the sediment surface eating indiscriminately anything small – probably all animals (many of which will themselves be juveniles) in a size range of 0.1–1.0 mm. Fortunately, there are some habitat types in brackish waters in which these predators do not appear to be able to exert such profound effects and which serve as refuges for susceptible prey species or life-history stages. The presence of surface relief in the form of shell debris or protruding worm tubes reduces the foraging success of mobile predators (Woodin, 1981), for example, and an even more important refuge system is provided by beds of submerged waterweeds and algae (Nelson *et al.*, 1982). It is presumably no accident that

Table 2. *Abundance of macrofauna (nos. per 0.1 m² retained on a 250 μm mesh sieve) in muddy sediments in the Waddenzee compared with that in areas enclosed, to exclude epibenthic predators, within 1 mm mesh cages for four months during late summer/ autumn*

	Uncaged mud	Caged mud
Hydrobia	2	10
Mya	0	87
Cerastoderma	7	1282
Spisula	0	185
Tubificoides	820	3055
Pygospio	17	350
Spio	2	50
Polydora	0	532
Malacoceros	0	405
Tharyx	7	5322
Capitella	92	140
Heteromastus	240	222
Nephtys	2	5
Eteone	0	117
Corophium	0	490
+ 14 other species	0	87
Total	1189	12339

Data from Reise (1978).

so many lagoonal species are associated with *Ruppia* and *Chaetomorpha*. Although information is still scarce, it may be that the numbers of the juvenile fish and crustaceans are themselves controlled by the larger predatory fish and birds, and that most of the epibenthic predators are themselves using the estuarine intertidal zone as a refuge from their own predators (Evans, 1983).

As we noted above there are also predators living within the sediments and this not only adds to (doubles?) the overall predatory impact but also can generate considerable complexity, not least in attempts to investigate the workings of the system. Thus, Kneib (1988) found that the small anemone *Nematostella* actually increased in the presence of a predatory fish. This counterintuitive situation resulted from one predator eating another. The fish were preying on the prawn *Palaemonetes* which was itself a consumer of the anemone. The position was in fact even more complex in that *Nematostella* is a predator on juvenile *Hydrobia* amongst others, so that increased survival of the anemone could lead to decreased survival of the mudsnail. Ambrose (1984) describes a similar *Glycera* /*Nereis* /'rest of the fauna' interaction in which the densities of most species are highest in the presence of increased numbers of the predatory bloodworm *Glycera* (which consumes *Nereis*) and lowest in the presence of elevated numbers of *Nereis*. There is some evidence to suggest that predatory worms within the sediments are indeed the preferred prey of many of the larger mobile fish and crustaceans, perhaps because many of the worms are active at or just below the sediment surface and are therefore more readily available to surface consumers and/or perhaps because of their greater average body size.

The effect of the predators mentioned above is not confined to the consumption of prey individuals, since in order to find food in soft sediments – in which most organisms live below the surface – it may be necessary to pick over the surface layers, to burrow, to excavate pits, to suck up mouthfuls of particles, and so on. These all create a disturbance to the habitat ('bioturbation'), equivalent to that of bait-diggers, that may uncover animals from concealment, destroy their burrow systems and generally destabilize the uppermost few centimetres of the sediments (Thistle, 1981). Juvenile fish may turn over every square metre of the sediment surface, down to a depth of 2 mm, whilst they are in their nursery area, bioturbating 1% of the available surface each day (Billheimer & Coull, 1988). At the very least this accidental disturbance might make animals that would otherwise be hidden within the sediment available to other consumers.

There are, however, several other processes at work in brackish waters besides predation that could limit the densities of the populations there. These include the effects of competition and of changes in the physical environment. Other factors such as alterations to the sediment caused by burrowing or feeding activities ('disturbance'), 'accidental' destruction of minute larvae or other early life-history stages by larger animals ('adult/larval interactions'), and so-called 'interference competition' between different organisms are not so much likely to set overall levels of abundance as to cause patchiness within distributions. They will be covered separately below.

The occurrence of competition for food or space in brackish waters is contentious. Levinton (1972) argued that deposit feeders ought as a rule to be food-limited and that competition between them should be intense. Suspension feeders, on the other hand, he considered theoretically unlikely to compete for food, although they could well be expected to compete for space. This hypothesis of the differential effect of food availability on deposit and suspension feeders has been tested by Olafsson (1986). The bivalve *Macoma* is one of those animals that can feed in more than one mode: it deposit feeds in summer and when in muddy sediments, but is predominantly a suspension feeder in winter and in sand (Zwarts & Wanink, 1989). Olafsson did indeed find evidence of strong competition in dense *Macoma* populations in muddy sediments, leading to reduction in individual growth rate, but no such effects in the sand-inhabiting individuals. In spite of this, most experimental studies in brackish waters have failed to find much evidence of competition in the field, either within or between species (although see, for example, the studies of Wilson, 1989, and Jensen & Kristensen, 1990, on competition in *Corophium* and the way in which this might be mediated by shorebird predation). Increasing the available food supply, for example, fails to result in larger invertebrate populations (Wiltse *et al.*, 1984).

Nevertheless, there is no doubt that the deposit-feeding burrowers in soft-sediments appear to have divided up their habitat between the available species. Some live in relatively sandy areas, others in mud; some live on the surface, others at depth; some live at low tidal levels, others near the top of the shore; and so on. Levinton interpreted this 'resource subdivision' as evidence of a legacy of competition in the past. Studies may not show that these species compete now, but unless they competed with each other in the past, how can we make sense of their present small-scale distributions? (It should be noted that this 'ghost of competition past' argument is untestable and as such should be an hypothesis of last resort.) Levinton's conundrum has other solutions, however; for example, avoidance of inappropriate matings, as in Kolding's studies on *Gammarus* described above, and some of the other causes of subdivision of the available habitat mentioned below. The inference of the lack of clear evidence for competition can only be that in the majority of areas and at most times the species concerned are being maintained below the carrying capacity of their habitat, in respect of both food and space, by other sources of mortality. It is then only when one or more populations evade these limitations – a rare event – that competition occurs (Barnes, 1994b).

It is abundantly clear, however, that the brackish-water environment can occasionally be extremely hostile. Storms, as we have seen, can have drastic results, and periods of low oxygen content of the water, together with catastrophically lowered salinities and reduced water levels, have been recorded as inducing mass mortality in Danish lagoons (Fenchel, 1975). In the Waddenzee, 'severe winters with prolonged frost [and ice] capable of killing all cockles' occur on average once every ten years (Reise, 1985). The *Cerastoderma* year-classes of 1976 and 1978 (the only two in a ten-year study period to escape devastation by the epibenthic predators) were both killed by severe winter conditions in early 1979, for example. Bivalves are not the only organisms to suffer during these Waddenzee 'ice winters'. Amongst other populations dramatically reduced are those of the predatory juvenile fish and crustaceans, and it is notable that recruitment of cockles is unusually high in the summers following widespread mortality of the consumers of their spat (Jensen & Jensen, 1985). Even, somewhat paradoxically, normal high spring tides can be a hostile element of the environment. Flooding tides roll large numbers of the mudsnail *Hydrobia* upshore, and high spring tides deposit vast carpets of this snail at the high water mark. Return downshore is impossible from there, since *Hydrobia* is active only when wet, and so they remain, stranded, until they die.

Unfortunately, no study has quantified the effects of all these different causes of mortality in any single estuary or lagoon, and so we really do not know what limits the numbers of brackish-water animals. On present evidence, it seems most probable that their numbers are usually kept well below the density that could be supported by the available food and space as a result of the action of predatory worms, crustaceans and young fish as well as by periods of environmental hostility, but further work is desperately needed. This is especially true in lagoons and brackish ponds, which are almost unstudied from these points of view. We do know rather more – although not nearly enough – about some of the interactions between different members of the fauna that can give rise to patchy distributions within the habitat as a whole.

A more covert form of predation than any described above is incidental destruction of larvae by adult animals of the same or of different species during their normal feeding or locomotory activities. The microscopic young are not necessarily consumed; they may simply be filtered from suspension in the water and then deposited onto the sediment bound up with inert particles in a mucous pellet, or they may be buried by sediment movements caused by larger organisms (Elmgren *et al.*, 1986). That passive destruction of juveniles is an important force determining the composition of shallow-water communities was pioneered by Woodin (1976), and since then negative links between successful larval settlement and resident gastropods, bivalves, polychaetes, tanaids, and amphipods have been shown. The cockle *Cerastoderma edule* and the clam *Mya arenaria*, for example, can reduce the numbers of settling bivalve spat by 40% and 20%, respectively (André & Rosenberg, 1991), and high densities of *Macoma* can halve the density of juvenile mudsnails (Olafsson, 1989). It has often been observed that all the bivalve molluscs of a certain species in one particular area are of the same age. This 'age–class effect' can be interpreted in terms of the destruction for several consecutive years of all the larvae attempting to settle by one successful year class of established adults. Eventually, on the death of this yearclass, the area is freed again for colonization by larvae that in turn may grow into adults that occupy the sediment exclusively for some time. Other explanations are possible, however, such as the effects of periodic 'ice winters' or the successful evasion of predation by one particular year group (see above).

Intolerance of the disturbance or habitat changes that other species can cause is frequent amongst sediment dwellers. In 1970, Rhoads and Young proposed an hypothesis in which deposit feeders rendered the habitat unsuitable for suspension feeders by making the surface layers of sediment fluid, full of faecal pellets and easily resuspended. Such soft muddy material would (a) clog the filters of the suspension feeders, (b) bury their newly settled larvae, and (c) cover any surfaces to which sessile species could otherwise attach. Since that time, further ecologically equivalent groups of species have been added to the hypothesis and its base has been widened from being purely a matter of feeding type to one embracing general life style as well (see, for example, Woodin & Jackson, 1979). Thus burrowing species, especially those ingesting sediment at depth and voiding it onto the surface or carrying out burrow construction, are responsible for considerable movement of sediment and consequent disturbance. Rates of deposition of up to 400 ml of wet mud per individual per year have been recorded for deposit feeders (Rhoads, 1974). Suspension feeders can have an equivalent or even greater effect, each individual depositing over 600 mg wet-weight of pseudofaecal and faecal material per day (pseudofaeces are the materials trapped by filter feeders but rejected, instead of being consumed); mussels in the Waddenzee have been found to deposit a layer of mud 60 cm thick in a two year period (Ehlers, 1988).

This movement of sediment can have marked repercussions. Wilson (1981), for example, recorded reduction in the abundance of the small worm *Pygospio* as a result of mortality brought about by lugworm burrowing, although other members of the fauna were not affected. Most commonly, it is the smaller species that are more severely affected; as they are, for example, by the movement of cockles in the Waddenzee as *Cerastoderma* ploughs through the surface sediment when changing feeding site (Jensen, 1985). The reaction of the disturbed species is often attempted emigration. *Pygospio*, for example, migrates away from areas subject to lugworm bioturbation, although lugworm populations can be extensive and there may be few refuges available for the smaller worm. The mechanical disturbance brought about by the ragworm *Nereis,* the cockle *Cerastoderma* or the lugworm *Arenicola* is also known to increase the emigration rate of *Corophium*. And when emigrating they are at greater risk from casual predation. In the German Waddenzee dense populations of *Corophium* are only found in areas of failed cockle recruitment. Animals are also capable of detecting subtle indications of the unsuitability of a given patch of sediment. The mudsnail *Hydrobia,* for example, can detect sediments that have previously been occupied either by itself or by *Corophium* (Morrisey, 1988) and take appropriate action; it can also distinguish between sediments rich and poor in its preferred diatoms and move so as to congregate in the former (Coles, 1979).

'Interference competition' is a category of competition-like effects that occurs in circumstances in which there is little or no reason to believe that any shared resource is really insufficient for the needs of all concerned. Intolerance of the presence of others appears to be the mechanism and not pre-emption of actual resources. Direct aggression often occurs, including, in *Nereis,* cannibalism (Kristensen, 1988). Both growth rate and survival of juvenile lugworms, for example, are negatively related to the abundance of ragworms. *Nereis* takes up residence within the burrow systems of the lugworms and not only interferes with their movements (necessary for respiration and defaecation), but bites the tail ends of the *Arenicola,* attacking them at frequent intervals and even killing them (Witte & Wilde, 1979). Ragworms will even leave their own burrows to harrass *Arenicola* living up to 15 cm away. The pieces bitten from the lugworms' rear ends are not normally eaten, indicating the aggressive rather than predatory nature of the behaviour. The end result is that the young lugworms migrate away from centres of ragworm interference. The ragworm also interacts with *Corophium* by a combination of aggressive predation, interference and disturbance (Olafsson & Persson, 1986; Rönn *et al.*, 1988). In most cases the displaced species have refuges in space or time – for example, areas outside the microhabitat range of the disturber or interferer – and hence the end result is a series of zones or patches each dominated by one species.

As a result of all these factors, and doubtless others as yet undiscovered (including the effects of parasites), not only are the distributions of brackish-water species dynamic within their environment as a whole, but individual populations may not occupy the same local patches of sediment from year to year. What is a dense patch of *Corophium* this year may be occupied by *Nereis* or *Arenicola* next year. No one has yet recorded in detail the differential occupation of a given local area over a long period of time, and there is great scope here for simple studies. There is also a need for more observations of interactions between pairs of species, including ones that seem well documented, because individual studies are still coming up with apparently contradictory results. Wilson (1981), for example, found that the burrowing activities of lugworms did not reduce *Corophium* numbers, yet Flach (1992) observed that *Arenicola* could cause the disappearance of even small populations of this amphipod. We need to know much more about the precise nature and circumstances of these effects before we can make generalizations that will withstand the test of time.

Reproduction

Adult animals are often able to tolerate a more extreme environment than are their young. In estuaries this finds expression in the seawards migrations of various crustaceans for the duration of the breeding season. The example *par excellence* is the introduced crab *Eriocheir* that can withstand fresh water when adult but which must reproduce in the ancestral sea; the native *Carcinus* also exhibits the same behaviour. Relatively few estuarine animals possess free-living larval stages and in those that do, e.g. *Arenicola*, *Nereis* and *Hydrobia ulvae* (though not the other hydrobiids) (Fig. 26), the larvae are often short lived or inhabit the surface layers of the sediment rather than being truly planktonic. Whether this is because of the problem of the ambient brackish water, however, is open to question, although passing all stages of development within the egg will isolate the sensitive young from the potentially unfavourable salinity regime. An alternative explanation relates to the underlying one-way (seawards) flow system in estuaries that is imposed by the fresh water input. Planktonic larvae would be swept out to sea away from the parental habitat and might therefore be disadvantageous. Another factor is historical accident. Amongst the dominant animal groups in estuaries (and lagoons) are the amphipods and isopods, and no member of either group, whether brackish-water or not, has a larval stage. The ancestors of the particular branch of the Crustacea to which both groups belong appear to have abandoned the larval stage tens or hundreds of million years ago. If the lack of a larval stage is advantageous in brackish waters, then these two groups are pre-adapted to life there, for reasons unconnected with the nature of the environment itself. A final explanation relates to size of the hatching young with respect to predation. By and large aquatic animals face an either/or choice: produce (a) many (but small) young that develop via a larval stage that feeds itself in the plankton or (b) few (large) young that can fuel their development with yolk provided by the mother and then hatch as small versions of the adult. Few small young is a possibility (albeit a somewhat unlikely one), but many large ones is not. If the young stages of brackish-water animals face severe predation from the juvenile fish, etc., then it will be advantageous to produce as large young as possible, so that on hatching they are already beyond the size range taken by at least some of the predators, and this rules out the larval stage.

There are two sets of observations of marine versus brackish-water reproduction that suggest that specific alterations to the breeding system have certainly evolved in brackish waters. First, a few gastropods occur in two reproductive forms (or possibly as pairs of sibling species differing only in life-history characteristics: see Munksgaard, 1990, in respect of *Rissostomia*), one in brackish, and often lagoonal, waters and the other in the sea; e.g. *Rissostomia membranacea* and *Brachystomia rissoides* (Rasmussen, 1944; Rehfeldt, 1968). In all such cases known, the brackish populations produce relatively few, relatively large eggs that hatch as small snails, whilst the marine ones lay more, smaller eggs that hatch into planktonic larvae. The second observation concerns the closely related pair of cockles *Cerastoderma glaucum* and *C. edule,* of which the first is lagoonal and the second is estuarine and, more typically, marine. In this pairing, breed-

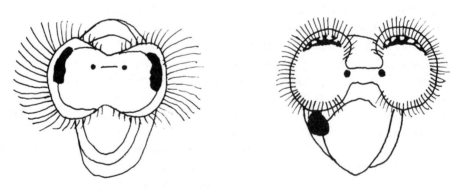

Eteone longa

Pygospio elegans

Nereis diversicolor

Hydrobia ulvae

Limapontia capitata

Fig. 26. The planktonic larvae of some brackish-water animals.

ing biology does not differ to any great extent, although *C. glaucum* may have shorter-lived larvae. The main difference between the two concerns the behaviour of the larvae at metamorphosis. Young lagoonal cockles inhabit the submerged waterweeds and (especially) filamentous algae, and only descend to the sediment when some millimetres in length. *C. edule,* in contrast, like 'normal' soft-sediment bivalves, settles onto the sediment surface at metamorphosis and burrows into it (Muus, 1967). We saw above that the submerged vegetation is an area of reduced predation relative to the unvegetated sediment. Circumstantial evidence therefore favours the anti-predator nature of the tendency towards loss of the larval stage, but there is no reason why all the various selective advantages outlined above should not have contributed to its suppression.

PART 2 Identification and notes on the species

Identification

In this section, all the larger animals (those retained by a mesh of 0.5 mm) to be found in the open water, sediment and submerged vegetation of lagoons, brackish ponds and similar pools will be keyed out, regardless of habitat salinity (i.e. including those lagoons in which the water is in fact full-strength sea water); but coverage of the estuarine fauna will be restricted to the larger animals (as above) of the intertidal mudflats in the permanently brackish zones. All keyed-out species are also illustrated (usually as whole animals) and an indication of their size is provided in the text (in the keys or in the notes, as appropriate).

The fauna of the estuarine water mass and subtidal zones will be omitted because of the inaccessibility of these areas and the difficulty of sampling them for most naturalists and students. Restriction to the 'permanently brackish zones' is necessary because of the problem mentioned on p. 2. Estuaries grade insensibly into fresh water and into the sea: near the mouths of estuaries (and estuarine lagoons) the fauna is essentially a fully marine one, whilst near the head it is almost entirely a depauperate fresh water one. Hence, to cover all the animals that might be found in an estuary this book would have to include the many marine species and a number of fresh water ones that can move or be swept into estuaries and the mouths of estuarine lagoons at certain states of the tide, and it would therefore have to be enormous. For estuaries then, coverage includes all those species to be found in the regions that do not normally experience salinities of less than 2‰ nor more than 27‰ (the estuarine brackish-water range as indicated by the analysis of Bulger *et al.*, 1993). Animals occurring in more dilute 'estuarine' waters than this range may be identified via Fitter & Manuel (1986), followed if appropriate by the relevant *Scientific Publication* of the Freshwater Biological Association, and essentially marine species can be identified from Hayward & Ryland (1990) and the *Synopsis of the British Fauna (New Series)* volumes of the Linnean Society of London / Estuarine and Coastal Sciences Association.

All aquatic habitats also grade insensibly into terrestrial ones, and hence there has to be some cut-off point along this axis too. It seems appropriate in a book on brackish waters to put the emphasis on aquatic species, and so the essentially terrestrial spiders, insects, etc. of the emergent sections of reed- and sedge-beds, and those of the surface and aerial parts of salt-marshes have been excluded.

Two groups of animals will not be keyed out to species. Brackish waters support numerous species of small oligochaete worms; for example at least 23 species of tubificids plus various enchytraeids in the southwest Netherlands alone (Verdonschot *et al.*, 1982), but their identification may require delicate dissection under a microscope and/or serial sectioning to provide details of the reproductive systems This is far from simple in

animals that may measure less than 20 mm in length! The reader for whom the identification of oligochaetes is a necessity or the professional ecologist is referred to Brinkhurst
(1982) for details of this group, which will here be keyed out only to family level.

The second block of species comprises the juvenile stages of certain insects, specifically waterboatman nymphs and the larvae of flies and beetles. Here there are also
problems of identification, but in this case the difficulties are either that the young
stages may all be effectively identical or that it is not always known to which species a
given larva belongs. The nymphs of waterboatmen resemble the adults but almost
invariably lack the structures used to diagnose the species. Fortunately, it is usually the
case that adults can also be found in any habitat supporting the nymphs, and identification should only be carried out on these adults; hemipteran nymphs will otherwise only
be keyed out to family. The brackish-water dipterans (flies) are only aquatic when larval. More than 50 types occur (many of them chironomids, but including dolichopodids, psychodids, ptychopterids, stratiomyiids, syrphids, tabanids, tipulids and ephydrids). To identify these, and the larvae of coleopterans (beetles), of which some 40 of
the species known from brackish waters as adults have aquatic young (all except
Helophorus and *Ochthebius*), it is necessary to keep them alive until they pupate and until
eventually an adult emerges. This may then be identified from the keys below (for the
beetles), from the *Handbooks* series of the Royal Entomological Society of London, or,
in the case of the chironomids, using Pinder (1978). (Cranston (1982) does provide a
key to chironomid larvae of the subfamily Orthocladiinae) In this guide, identification
of fly larvae will proceed only to family level and of beetle larvae to genus.

In order to avoid harming the animals during identification, the keys below are
based on features visible in living animals under a low-power stereomicroscope (×6 –
×15); indeed, a number of them are based on features visible *only* in living material. In
some cases, especially in respect of arthropods, it will prove necessary to quieten the
animal in order to see the diagnostic features. This can be achieved by anaesthetization:
adding a small globule (< 1% by volume) of propylene phenoxetol to the medium for
species obtaining their oxygen from the water (available from Nipa Laboratories,
Treforest, Wales); or 'spraying' with carbon dioxide gas, as from a waterless Soda
Water Siphon or equivalent system, for air-breathing forms such as beetles – followed
in either case after identification by rapid return to a large volume of their natural
medium. It follows from the above, that the reader is strongly encouraged (a) to treat
brackish habitats with respect (many have protected status within nature reserves and
are subject to local conditions of access and/or bye-laws) and (b) to maintain collected
animals in a healthy state and to return them to their water body and precise microhabitat of origin after examination and identification. This latter point is not just for the
sake of the animals. Many populations, especially in land-locked systems, are small and
isolated; they are therefore of great potential interest to geneticists studying evolution.
If, however, animals are taken from one site and released into another, gene pools will
be mixed and evolutionary differences accumulated over generations may be weakened.
Clearly, it is important for the biogeographical, evolutionary and other study of brackish waters that deliberate introductions should not occur, or should not occur unless
authorized and recorded.

The reader is also reminded that permission may be necessary to gain access to, and to take samples from, a given site, and that some brackish-water species (the stonewort *Lamprothamnium papulosum,* the sea-anemones *Edwardsia ivelli* and *Nematostella vectensis,* the amphipod *Gammarus insensibilis,* the polychaete *Armandia cirrhosa* and the bryozoan *Victorella pavida*) are protected under Schedules 5 and 8 of the Wildlife and Countryside Act 1981. Since 1988, it has been illegal in Britain to catch or handle, let alone kill, any of these species without a specific licence from the national Nature Conservation agency. Admittedly, this places the field worker in an almost impossible situation that, strictly speaking, can only be resolved by never attempting to catch anything. It arises because none of the animals (all in the list except *Lamprothamnium*) can be recognised for what they are in the field: all are small, difficult to identify, and can easily be over-looked or mistaken for something else. In other words, one does not know that it has been collected until it is identified as such, and by then an offence has already been committed. The understandable result of this would be for the collector to keep very quiet indeed! This would be a great pity, because three of the species are each only known from a single site and therefore the finding of more localities (of any of the six species) would be a matter of great conservation significance. In the U.K. all are restricted to England, and if (of course 'by accident') the reader does find any of the listed species, or any others stated to be rare below, they are urged to contact English Nature, Northminster House, Peterborough, PE1 1UA (or should they be worried about any legal consequences of their discovery, they can feel free to contact this author, in confidence). A further nine species have British Red Data Book status: the amphipod *Corophium lacustre,* the gastropods *Pseudamnicola confusa, Truncatella subcylindrica, Paludinella littorina, Caecum armoricum* and *Tenellia adspersa,* and the insects *Sigara striata, Berosus spinosus* and *Paracyamus aeneus* (Shirt, 1987; Bratton, 1991).

Finally, readers should note that the keys that follow use features seen in the brack-ish-water members of the various species, genera and other taxa covered. These fea-tures are not necessarily also found in fresh water and marine members of the same groups, and hence the keys may not work in non-brackish habitats. Common names have been given below in cases where the animals genuinely have common names. All too often, however, the alleged common names of invertebrates are entirely spurious; sometimes no more than translations of the scientific name. I have avoided these, together with such abominations as the 'laver spire-shell' and the 'peppery furrow-shell': the animals concerned are indeed common, but the populace-at-large has yet to bestow recognition on them. As Wilson (1992) has pointed out, such only happens if there is some practical need to do so, and perhaps not surprisingly estuarine mudflats seem to have been avoided by the large majority of people. The other dictionary defini-tion of 'brackish', besides the one given at the beginning of this book, is after all 'nau-seous'!

Key to major groups of brackish-water animals

Those readers able to recognize the various major groups of aquatic animals can proceed straight to the relevant key to species in the following pages.

1 A Animals with all or part of their bodies visibly in the form of a linear chain of segments (n.b. some segments may be larger than others) 2

   B Animals with unsegmented bodies, or appearing to be unsegmented 13

2 A Body with series of 8 shell plates along upper surface, otherwise unsegmented .... **Chitons** (p. 105)
   [Slug-like animals, without legs, that are only apparently segmented. The region of the 8 shell plates is surrounded by a tough, unsegmented area of 'skin'. Cling to hard surfaces by a limpet-like foot.]

   B Body not as above 3

3 A Body tadpole-shaped, with a large pointed anterior portion formed by a carapace plus 5 free segments, from which issues a long, thin 'tail' of 6 segments that ends in a spiky trifid tail fan; the anterior portion bears 5 pairs of legs .... **Cumaceans** (p. 162)
   [Small crustaceans, < 20 mm long, inhabiting soft sediments. The first 2 pairs of legs are directed forwards and the following 3 pairs, which are used for burrowing, are directed backwards.]

   B Not as above 4

4 A Body covered by a hard exoskeleton, with or without jointed legs 5

   B Body soft; if legs present then only 3 pairs 11

5 A With legs 6

   B Without any legs .... **Insect larvae** (p. 206)
   [Some fly larvae possess relatively hard cuticles and hence might key out here]

6 A With 3 pairs of legs .... **Insects** (p. 210)
   [Some adult insects may appear unsegmented – except for their possession of a head, thorax and abdomen – when viewed from above, but their segmental nature is evident on the under surface. The nymphs of waterboatmen are more obviously segmented and differ from the adults externally only in their size, lack of wings and absence of sexual characters.]

   B With 5-7 pairs of legs 7

7 A With 7 (rarely 5) pairs of uniramous walking legs that are all effectively identical, except for size, and that each end in a simple point .... **Isopods** (p. 162)
   [Woodlouse-like animals that are flattened from top to bottom and that crawl across the bottom (or, rarely, swim). Without a carapace; with 5 pairs of pleopods (sometimes hidden) and 1 pair of uropods.]

   B Not as above 8

8 A With 7 pairs of uniramous walking legs, of which the first two have enlarged ends and are nearly always subchelate; without a carapace .... **Amphipods** (p. 170)
   [Crustaceans that are flattened from side to side, and that each possess walking legs of a variety of types. With 3 pairs of pleopods and 3 pairs of uropods.]

   B Not as above 9

9 A With 7 pairs of uniramous walking legs of which the first is large and ends in a chela; with a small carapace .... **Tanaids** (p. 168)
[Small, elongate crustaceans with a carapace from beneath which emerges the cheliped, the remaining thoracic segments being separate and each bearing a pair of uniform, uniramous walking legs. Bottom dwelling.]

   B Not as above 10

10 A With 5 (or apparently 6) pairs of uniramous walking legs of which the first 1 or 2 pairs (or nos 2 and 3) end in a chela; with a single large carapace .... **Decapod crustaceans** (p. 200)
[Swimming or crawling crustaceans with a carapace covering the body region from which emerge the uniramous walking legs. In the crabs, the carapace may cover the whole body when viewed from above, the segmented abdomen being tucked under the body. In some of the swimming species (the prawns), the last pair of mouthparts may be large and resemble a walking leg.]

   B With 5-7 pairs of biramous swimming legs; with a two-part carapace .... **Mysids** (p. 156)
[Fragile, swimming, shrimp-like crustaceans with a two-part carapace partially covering the body region from which the legs emerge (but exposing the upper surfaces of the last two thoracic segments). No legs end in chelae.]

11 A Body with a head and 12 or less apparent segments .... **Insects** (p. 206)
[Insect larvae vary from being soft bodied and limbless, to possessing 3 pairs of jointed legs and/or many unjointed projections from their segments and having hard plates on some segments.]

   B Body with more than 14 segments or, if with less, then with an obvious system of tentacles arising from the head 12

12 A Segments without any projecting, chaeta-bearing flanges, lobes or paddles, and without any tentacles or equivalent projections, except in one species a pair of suckers, one at each end of the body .... **Clitellate annelids** (p. 102)
[Either small versions of the familiar earthworms that, like the latter, usually bear a clitellum in the breeding season, or leeches. Always with more than 15 segments.]

   B At least some segments with projecting, chaeta-bearing flanges or larger outgrowths, and often with tentacle-like projections and eyes; nearly always with many more than 15 segments, but 3 small species possess fewer and they have crowns of tentacles
.... **Polychaete annelids** (p. 78)
[The bristle-worms exhibit a wide range of body form from free-living to burrow- or tube-dwelling, from possessing a uniform series of segments to the differentiation of various body regions, and from bearing many appendages to having these reduced down to a pair of lateral, chaeta-bearing flanges on some segments.]

13 A Animals forming a colony (i.e. more or less similar individuals in tissue contact with each other or jointly inhabiting a single, secreted colonial structure) 14

   B Non-colonial, solitary animals 17

14 A Colonies in the form of an encrusting sheet, plate, bush or mound of calcareous or gelatinous boxes, each box being some 0.5 mm (and always < 1 mm) in largest dimension
.... **Bryozoans** (p. 244)
[Each box contains one colonial 'individal' (zooid) many or all of which can protrude a small circular series of tentacles out of an aperture in the box and withdraw them completely within for safety. Retracted zooids are usually visible through the roof of the box.]

   B Not as above 15

15 A Colonies in the form of brightly coloured jelly in which the individuals are set in star-shaped or zip-fastener patterns .... **Ascidians** (p. 252)
[Encrusting mounds or attached lobes formed by colonial sea-squirts in which the individual zooids are 2–3 mm in largest dimension. Each zooid has its own inhalent siphon but discharges through common exhalent openings in the colony.]

B Colonies in the form of a creeping stolon from which arise single or groups of individual tentaculate organisms at intervals, or in the form of a miniature bush bearing many such individuals along its branches.. 16

16 A No part of the colony covered by a thin, horny, protective, external casing
.... **Bryozoans** (p. 244)
[Each cylindrical 'tube' contains one colonial 'individal' (zooid) that can protrude a small circular or horseshoe-shaped series of tentacles out of the terminal aperture of the cylinder and withdraw them completely within for safety. Retracted zooids are usually visible through the walls.]

B At least part, and sometimes all, of the colony within a thin, horny, protective casing
.... **Cnidarians** (p. 56)
[Some hydroids, those in which the protective casing extends in the form of a cup around each tentaculate polyp, possess polyps with a circlet of tentacles around the mouth; those in which the protective casing does not extend around the polyp generally (in brackish waters) possess polyps with tentacles in more than one whorl or distributed more generally over the polyp surface. Some polyps lack tentacles and are vase-shaped or bulbous and specialized for reproduction.]

17 A With body wholly or partly enclosed within a hard shell 18
B Body not wholly or partly enclosed within a hard shell 22

18 A Shell a flat cone formed by 4–6 radial plates with 4 small plates at the apex (centre)
.... **Barnacles** (p. 154)
[Barnacles are really segmented animals, but their segmentation is not externally visible. Small feathery (feeding) legs may be seen to emerge from the apex of the shell.]

B Body not as above 19

19 A Body wholly or partly enclosed within a bivalved shell 20
B Shell not in the form of two hinged plates 21

20 A Small (< 1 mm) swimming animals .... **Ostracods** (p. 156)
[Ostracods are really segmented animals, but their segmentation is not externally visible. 2 or 3 pairs of small appendages may be seen to extend beyond the confines of the shell when the animals move.]

B Small to large, sedentary or sessile animals .... **Bivalve molluscs** (p. 138)
[A pair of siphons (sometimes joined together) and/ or a muscular foot may protrude from between the two valves of the shell, but brackish-water species do not swim and are immobile in comparison with the ostracods, the only other animals with a bivalved shell]

21 A Animals wholly or partly enclosed with a single, coiled or cap-shaped shell
.... **Shelled gastropod molluscs** (p. 105)
[When moving, the shell is usually carried in the middle of the body between the anterior head (bearing eyes and tentacles) and the posterior tail end of the foot; in limpet-shaped species, however, neither head nor foot may be visible during locomotion, and in some sea-slugs a shell may be present but partially concealed by soft tissue.]

B Body with a series of 8 shell plates along upper surface .... **Chitons** (p. 105)
[Slug-like animals, without legs. The region of the 8 shell plates is surrounded by a tough, unsegmented area of 'skin'. Cling to hard surfaces by a limpet-like foot.]

22 A Soft-bodied animals with a circlet or circlets of tentacles around the mouth
.... **Cnidarians** (p. 56)
[Sea-anemones may be squat and attached to hard surfaces (sometimes stones or shells buried in the sediment) or be more elongate and buried in sand or mud. All will extend their tentacles if undisturbed. Jellyfish are somewhat equivalent to flattened, upsidedown sea-anemones and they also key out here.]

B Not as above 23

23 A Leathery attached animals with 2 permanent siphons............................. **Ascidians** (p. 252)

   [*Solitary sea-squirts are sessile, blob-like animals attached to stones, vegetation, etc., sometimes by means of a stalk, that have 2 apertures to their body, out of which water squirts if gently squeezed.*]

   B Not as above 24

24 A Animals with 1 or more fins along the dorsal and ventral midlines and around the tail end; head with an obvious mouth and pair of large eyes...................................... **Fish** (p. 254)

   [*Powerful mobile swimmers often with 1 or 2 pairs of paired fins in addition to the median ones.*]

   B Animals other than fish 25

25 A Soft bodied animals 26

   B Body hard, although not covered by a shell; flat animals with a small central disc from which radiate 5 long arms... **Ophiuroids** (p. 250)

   [*Brittlestars with a small central body containing a mouth in the middle of the lower surface, and five long, sinuous and bristly arms.*]

26 A Slug-shaped animals... **Shell-less gastropod molluscs** (p. 112)

   [*The solid-bodied sea-slugs crawl on a large foot that occupies the lower surface, and usually possess a variety of projections from the upper surface. Many are brightly coloured. Some possess a shell that is hidden beneath body tissue. Most often found in association with their hydroid or bryozoan prey.*]

   B Worms... 27

27 A Small, stiff worms, pointed at each end, that move in C- or S-shaped wriggles
   .... **Nematodes** (p. 76)

   [*Translucent worms that are round in cross section. With 1 exception too small to be collected by 500 µm sieves. Without eyes.*]

   B Not as above 28

28 A Sausage-shaped worms with an anterior region that can be retracted into the body or slowly unrolled when feeding.. **Sipunculans** (p. 104)

   [*Opaque worms that are round in section and that extend and retract a long thin anterior portion tipped by fleshy lobes during feeding. Sedentary, slow moving and without eyes.*]

   B Not as above 29

29 A Flat worms that glide along surfaces without changing shape
   .... **Flatworms** and **Nemertines** (p. 70)

   [*Appendage-less and often highly pigmented worms, sometimes of great length but of little height or breadth, that may contract when disturbed but otherwise move smoothly without wriggling or any change in length or breadth. Many possess eye spots.*]

   B Small 'worms' attached at one end to sand grains...................... ***Protohydra leuckarti*** (p. 60)

   [*This solitary hydroid, measuring < 2 mm in length, is unusual amongst cnidarians in lacking tentacles around its mouth, and particularly when extended it most closely resembles a worm. It has a knobbly skin, is largest away from its point of attachment, and is often bright red in colour.*]

Cnidarians (hydroids, sea-anemones and medusae)

With the exception of the jellyfish *Aurelia aurita* and a few medusae, brackish-water cnidarians are solitary or colonial, microscopic to 50 cm tall, sea-anemone-like animals that are attached to objects or shallowly buried in the sediment. In these, tentacles (absent in one species) occur in a circle or more diffusely around the mouth region, whilst the other end of the body forms an attachment disc, a bulbous burrowing organ, or a small tube joining those of other members of the colony. Colonial species are wholly or partly encased in a thin, tubular, protective covering. The medusae are all saucer- or bell-shaped, transparent members of the plankton, with four or more tentacles dangling from their circular lower margin. Manuel (1988) describes the British sea-anemones. See Fig. 27.

Key to Benthic Cnidarians (Polyps)

1 A Animals solitary, attached to hard surfaces or partially buried in soft sediment 19
  B Animals forming modular colonies of individual polyps attached to each other and situated on branches of a communal bush-like colony or arising from a communal stolon; colonies attached to firm surfaces (stones, wood, vegetation, mollusc shells, etc.) 2

2 A Feeding polyp with tentacles scattered irregularly over its external surface 3
  B Feeding polyps with tentacles arranged in one or two distinct and discrete whorls 7

3 A Tentacles of feeding polyps with club-tipped or bulbous ends 4
  B Tentacles of feeding polyps not swollen at their ends 5

4 A Feeding polyps arising singly from the communal stolon .... *Sarsia tubulosa*
  B Several feeding polyps occur on branched stalks.. *Sarsia loveni*

5 A Feeding polyps arise directly from the basal stolon or encrusting mass .... *Clava multicornis*
  B Feeding polyps borne on a stem arising from the stolon 6

6 A Feeding polyp with tentacles confined to its distal region .... *Clavopsella navis*
  B Feeding polyp with tentacles scattered over its whole surface..................... *Cordylophora caspia*

7 A Tentacles of feeding polyps clearly in two whorls (one, oral, whorl of small tentacles around the mouth, plus a second, aboral, whorl of much larger ones) 8
  B Tentacles of feeding polyps in only a single whorl 9

8 A With 14–20 oral and 20 aboral tentacles .... *Tubularia larynx*
  B With 40 oral and 20–30 aboral tentacles .... *Tubularia indivisa*

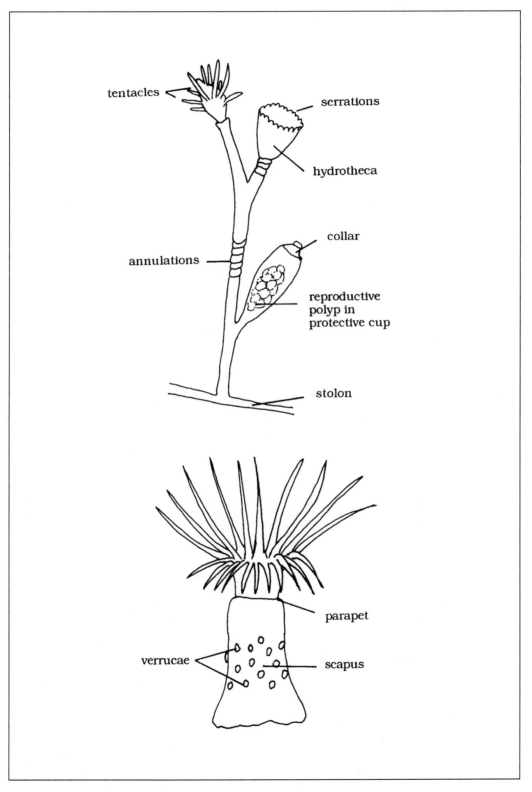

Fig. 27. Anatomical features used in the identification of cnidarians.

9 A Feeding polyps arising singly from a stolon .... *Halitholus cirratus*

 B Feeding polyps on stem or branches of ascending parts of the colony 10

10 A Feeding polyps mounted directly on the main communal stem, not on stalks 11

 B Feeding polyps on stalks 12

11 A Feeding polyps issuing alternately from the stem .... *Sertularia cupressina*

 B Feeding polyps issuing in opposite pairs .... *Dynamena pumila*

12 A Stalks of feeding polyps clearly annulate in places 13

 B Stalks of feeding polyps not ringed 18

13 A Protective cup (hydrotheca) around feeding polyp with smooth, non-serrated rim
 .... *Laomedea flexuosa*

 B Hydrotheca with serrated rim 14

14 A Serrations (cusps) on hydrothecae themselves serrate (n.b. the secondary serrations may be
 worn away in old hydrothecae)... 15

 B Serrations on hydrothecae simple, without further serrations.. 16

15 A Colony bushy, up to 35 cm tall, arising from dense mesh of stolon threads; hydrotheca
 with 12–4 cusps.. *Hartlaubella gelatinosa*

 B Colony slender, up to 15 cm tall; hydrotheca with 10–20 cusps .... *Obelia bidentata*

16 A Protective cup around reproductive polyp with terminal collar; reproductive polyps bud off
 medusae within the cups; colonies arise singly from the substratum 17

 B Protective cup around reproductive polyp without terminal collar; when mature, vestigial
 attached medusae project in twos or threes from the cups; small slender colonies arise
 from creeping stolon .... *Gonothyraea loveni*

17 A Colony elongate, ± parallel sided; older regions of stem black .... *Obelia longissima*

 B Colony fan-shaped; older regions of stem not black .... *Obelia dichotoma*

18 A Feeding polyps borne on one side of the colonial branches only .... *Ventromma halecioides*

 B Feeding polyps do not arise only from one side of the colonial branches .... *Bimeria franciscana*
 [This species has often been recorded as Perigonimus megas *or* Garveia franciscana.*]*

19 A Animal not attached to any structure but freely burrowing in soft sediments...................... 20

 B Animal with one end fixed to a solid object (rock, stone, shell, or in one case sand grains)
 by a basal attachment disc, although the solid object may itself be buried in soft sediments
 21

20 A Body partly covered by a thin, translucent outer cuticle; burrows in soft mud; with 12 tentacles spotted in brown or cream .... *Edwardsia ivelli*
 [This species has not been seen for a number of years in the only locality from which it was known.]

 B Without outer cuticular skin, although often with covering of mucus and debris; amongst
 filamentous algae or associated with soft sediments; usually with 14–16 very long, white-
 banded tentacles .... *Nematostella vectensis*

21 A Animal larger than 2 mm, with feeding tentacles around the mouth; body attached to rock, stone or shell 22

B Animals without tentacles; small (< 2 mm), worm-shaped, warty body attached to sand grains .... *Protohydra leuckarti*

22 A Tentacle number in excess of 500 .... *Cereus pedunculatus*
B Tentacle number 200 or less.. 23

23 A Body differentiated into two regions: a large, thick-walled, basal region (the scapus), separated from a smaller, thinner-walled area that bears the tentacles by a parapet (an infolded rim of the scapus enclosing a distinct groove) 24
B Body not differentiated into two regions 27

24 A Scapus with projecting wart-like verrucae, to which sand and debris adhere, scattered over its surface; tentacles short and stout .... *Urticina felina*
B Scapus without scattered warts; tentacles not short and stout 25

25 A Brown to green scapus with vertical orange (sometimes paler) stripes .... *Haliplanella lineata*
B Scapus without vertical orange stripes 26

26 A Body, including tentacles, orange; without prominent warts in the groove within the parapet .... *Diadumene cincta*
B With row of (usually blue) warts in the groove within the parapet; body and tentacles not usually orange .... *Actinia equina*

27 A Tentacles arranged irregularly .... *Sagartia elegans*
B Tentacles arranged in regular whorls, in multiples of six... 28

28 A Base up to 50 mm diameter; with up to 192 tentacles; in nature, usually with sediment and debris adhering to suckers on the upper part of the body .... *Sagartia troglodytes*
B Base up to 15 mm diameter; with up to 96 tentacles; usually without sediment or gravel adhering to the suckers; obligately parthenogenetic...................................... *Sagartia ornata*
[These two species are difficult to separate on a morphological basis.]

Key to Planktonic Cnidarians (Medusae)

1 A Medusa saucer-shaped, with numerous (24 or more) marginal tentacles 2
B Medusa bell-shaped, with < 24 individual or clumps of marginal tentacles 3

2 A Medusa small (< 6 mm diameter), with tentacles of length at least half the diameter of the disc; mouth without obvious lobes .... *Obelia* spp.
B Medusa up to 250 mm (or more) diameter, with tentacles only one twelfth or less diameter of disc; mouth with four large lobes almost equal to disc radius in length .... *Aurelia aurita*

3 A With four marginal tentacles and median lobe bearing the mouth extending well beyond margin of bell (up to 16 mm diameter) .... *Sarsia tubulosa*

   B With eight or more marginal tentacles; median lobe not extending beyond margin of bell 4

4 A With eight clumps of marginal tentacles, each clump with up to three individual tentacles; up to 4 mm diameter .... *Rathkea octopunctata*

   B With some 22 marginal tentacles evenly distributed around the margin of the bell; up to 14 mm diameter .... *Halitholus cirratus*

HYDROZOA
Athecata
Hydridae
Protohydra leuckarti (Fig. 28)
A small (1–2 mm long), solitary, tentacle-less hydroid widespread not only in northwest Europe but throughout the northern hemisphere, although rarely recorded as a result of its small size and non-descript appearance. It has been recorded over a salinity range of 4–30‰, in soft sediments, and at densities of up to 200 000 individuals/m^2. Its colour is derived from its food: red when predominantly consuming interstitial copepods, and colourless on a diet of nematodes. Multiplication is by transverse division, and movement by peristalsis. No free-living medusoid stage occurs.

Tubulariidae
Tubularia indivisa (Fig. 28); *Tubularia larynx* (Fig. 28)
These *Tubularia* spp. are able to withstand some dilution of their environment and have been recorded from piers, etc. in the mouths of estuaries. *T. indivisa* may exceed 150 mm height, but *T. larynx* is <50 mm. No free-living medusoid stage occurs.

Corynidae
Sarsia tubulosa (Fig. 28); *Sarsia loveni* (Fig. 28)
Both these widely distributed hydroids can penetrate into brackish waters and have been recorded from estuaries and lochs on submerged vegetation and hard surfaces. They rarely occur in non-tidal conditions, however. Up to 40 mm height. Only *S. tubulosa* possesses a free-swimming medusa.

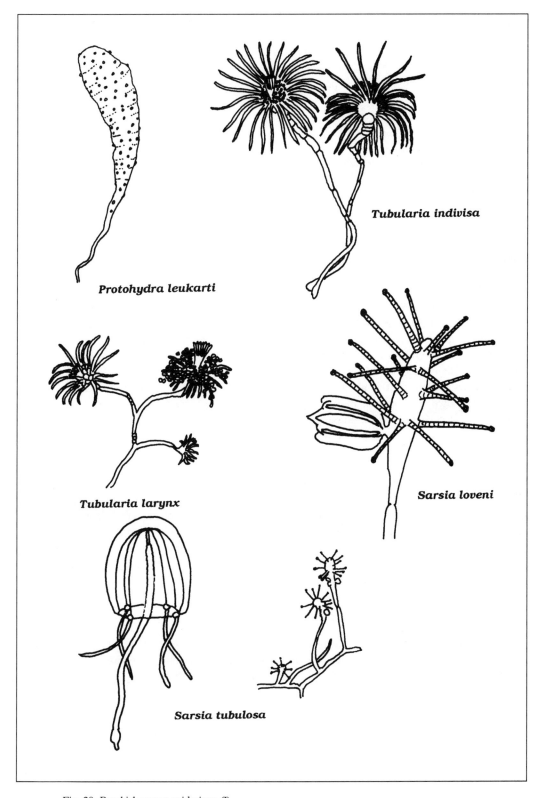

Protohydra leukarti

Tubularia indivisa

Tubularia larynx

Sarsia loveni

Sarsia tubulosa

Fig. 28. Brackish-water cnidarians (I).

Clavidae

Cordylophora caspia (Fig. 29)

Probably essentially a fresh water species that can colonize brackish waters up to salinities of over 20‰ (and occasionally to 35‰) throughout the region. Colonies grow on reeds, submerged vegetation, hard substrata such as wooden posts, and have even been found growing on the shells of living crabs and gastropods. A salinity optimum of 17‰ has been suggested. No free-living medusoid stage occurs. Up to 100 mm height.

Clava multicornis (Fig. 29)

A widespread hydroid on intertidal algae over the seaward halves of estuaries. No free-living medusoid stage occurs. Up to 30 mm high.

Rathkeidae

Rathkea octopunctata (Fig. 29)

The hydroid stage is minute and most unlikely to be encountered, but the medusa is up to 4 mm diameter and is widely distributed in lagoons and other brackish waters throughout the region.

Bougainvilliidae

Bimeria franciscana (Fig. 29)

This hydroid, which has been widely recorded from Belgium to Germany on hard substrata in nontidal ponds and ditches, and to a lesser extent estuaries (Schelde to Elbe), appears in the literature under the names *Garveia franciscana* (Dutch records) and *Perigonimus megas* (German records). It can grow over a salinity range of 3.5–35‰, and can survive down to 1.5‰. It was particularly abundant in the former Zuiderzee. It is almost certainly an alien species accidentally introduced from elsewhere: it is known from Africa, India, Australia, both coasts of the U.S.A. and the Mediterranean (from the Venice Lagoon, for example), but some of these could also be introductions. No free-living medusoid stage occurs. Up to 150 mm high.

Clavopsella navis (Fig. 29)

A small species known only from the Kiel Canal, Germany (under the name *C. quadranularia*), from Widewater Lagoon, Sussex, and outside northwest Europe from the Azores and from a ship's hull in South Africa (under the generic name *Rhizorhagium*). In Widewater, in which it is abundant, it grows on algae such as *Chaetomorpha*. Its lower level of salinity tolerance would appear to be in the order of 8‰. The systematic position of the genus *Clavopsella* is uncertain: at one extreme, it has been regarded as constituting a separate family, and at the other of being synonymous with *Cordylophora*. No free-living medusoid stage occurs. Up to 30 mm high.

Pandeidae

Halitholus cirratus (Fig. 29)

A small hydroid based in the Baltic Sea that occurs in the extreme east of the region in Denmark. For many years the polyp was known under the generic name *Perigonimus* whilst the medusa was termed *Halitholus*. Hydroid up to 3 mm high; medusa up to 14 mm diameter.

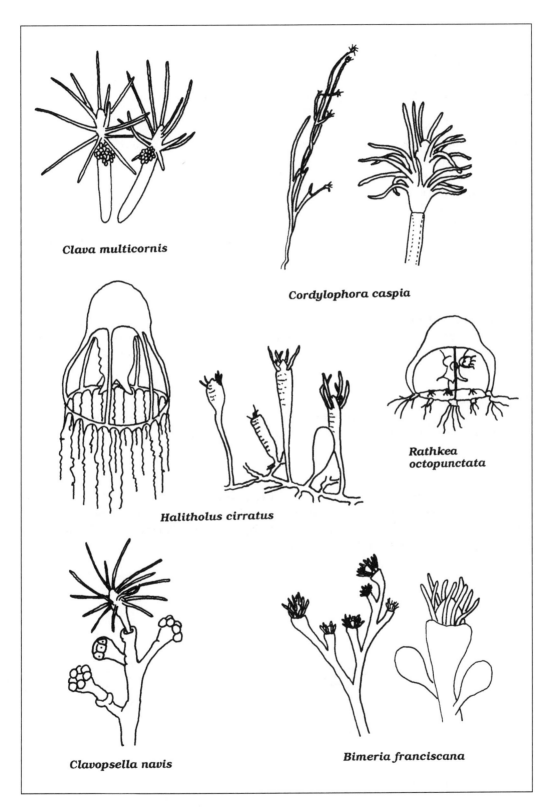

Fig. 29. Brackish-water cnidarians (II).

Thecata

Sertulariidae

Dynamena pumila (Fig. 30)

Occurs on intertidal algae and hard substrata over the seaward portions of those northwest European estuaries with rocky outcrops. No free-living medusoid stage occurs. Up to 30 mm high.

Sertularia cupressina (Fig. 30)

Rarely recorded from the mouths of estuaries. No free-living medusoid stage occurs. Up to 50 cm tall.

Plumulariidae

Ventromma halecioides (Fig. 30)

An almost cosmopolitan species that has occasionally been recorded growing on *Zostera* and hard substrata in relatively high-salinity lagoons in the region. No free-living medusoid stage occurs. Up to 75 mm tall.

Campanulariidae

Gonothyraea loveni (Fig. 30)

Often recorded under the generic name *Laomedea* , this hydroid is abundant in many lagoons and equivalent habitats growing on *Ruppia* and other hard substrata in salinities of down to some 8‰. It also occurs in estuaries and sheltered marine waters throughout the region. No free-living medusoid stage occurs. Up to 30 mm tall.

Laomedea flexuosa (Fig. 30)

A very common hydroid throughout Europe that occurs in estuarine and lagoonal brackish waters as well as in the sea. No free-living medusoid stage occurs. Up to 35 mm tall.

Obelia dichotoma (Fig. 30); *Obelia longissima* (Fig. 30); *Obelia bidentata* (Fig. 31); *Hartlaubella gelatinosa* (Fig. 31)

These current and former species of *Obelia* are all tolerant of a degree of brackish water and can occur in the more saline regions of estuaries and non-tidal northwest European waters, growing on submerged vegetation and/or posts and stones. Usually up to 350 mm long. The three *Obelia* spp., but not *H. gelatinosa*, possess free-living medusae.

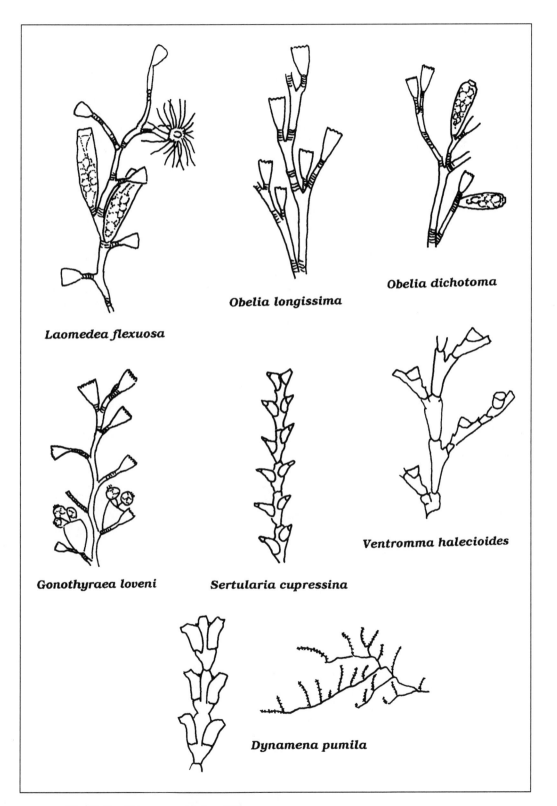

Fig. 30. Brackish-water cnidarians (III).

SCYPHOZOA
Aureliidae

Aurelia aurita (moon jelly) (Fig. 31)

A cosmopolitan species that is known in the Baltic down to salinities of 3‰, and can breed there down to 6‰, and elsewhere sporadically occurs in large numbers in estuaries, brackish lochs and harbours though never to such low salinities. It is also known from those few lagoons that have a direct connection to the sea (including via overtopping). Some local estuarine populations may be distinct. Diameter up to 25 cm.

ANTHOZOA
Actiniaria
Actiniidae

Actinia equina (beadlet anemone) (Fig. 31)

This common anemone is mainly a rocky-shore species, but it can withstand considerable dilution of sea water (down to near 8‰) and it occurs in the mouths of estuaries, in areas of fresh water discharge on beaches, etc., wherever suitable rocks for attachment are present. Linnaeus originally described this species as *Priapus equinus:* presumably his Swedish specimens were rather larger than those with which the author is familiar! Base diameter up to 50 mm.

Urticina felina (dahlia anemone) (Fig. 31)

This abundant, circum-Arctic, rocky-shore species, for many years known as *Tealia*, occurs occasionally in gravel and in mussel beds in brackish water. It is common in the Kattegat, but is not notably an estuarine or lagoonal species. Base may be >100 mm diameter.

Diadumenidae

Haliplanella lineata (Fig. 31)

Originally introduced some 100 years ago, *H. lineata* is now a common brackish-water anemone throughout the region, being found attached to any suitable surface in lagoons, creeks, harbours and estuaries. It only rarely occurs in the sea. Also known as *Diadumene luciae*. Base <25 mm diameter.

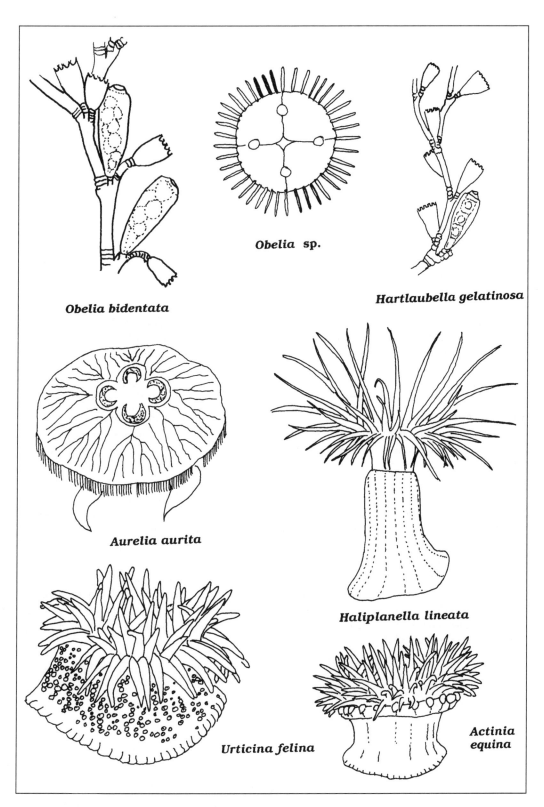

Obelia sp.

Obelia bidentata

Hartlaubella gelatinosa

Aurelia aurita

Haliplanella lineata

Urticina felina

Actinia equina

Fig. 31. Brackish-water cnidarians (IV).

Diadumene cincta (Fig. 32)

A northeast Atlantic species that is especially characteristic of brackish water, occurring in estuaries, mussel beds, and non-tidal lagoons throughout the region, as well as on rocky marine shores (especially sublittorally). Base diameter up to 10 mm.

Sagartiidae

Sagartia troglodytes (Fig. 32) and *Sagartia ornata* (Fig. 32)

These two species were until recently considered to be but varieties of the one *S. troglodytes*. Both occur in a wide variety of coastal habitats, including being buried in soft sediments and *Zostera* beds (in reality attached to a buried shell, stone, etc.) in lagoons, creeks and estuaries. *S. troglodytes* (*sensu stricto*) is the more frequently encountered species in such brackish areas.

Sagartia elegans (Fig. 32)

Although mainly an open-coast, rocky-shore species, *S. elegans* does uncommonly occur buried in soft sediments with the other northwest European sagartiids in the more saline of brackish-water regions. Base diameter up to 30 mm.

Cereus pedunculatus (Fig. 32)

Sometimes abundant in estuarine mudflats in the southwest of the region (only very rarely in the North Sea), its disc flush with the mud surface and its base attached to a buried stone or shell. Found only in the higher salinity reaches. The disc may achieve 150 mm diameter.

Edwardsiidae

Nematostella vectensis (Fig. 32)

An easily overlooked little species that burrows, often abundantly, in muds or sandy muds, or lives entwined in filamentous algae such as *Chaetomorpha* or associated with the larger *Ruppia* or *Zostera*, in non-tidal lagoons and, less frequently, in salt pans down to salinities of 8‰. Animals such as *Hydrobia* spp. and chironomid larvae (often larger than the anemone itself) have been observed in its gut; indeed most species, including quite large *Nereis* can be captured and killed, but their size may prevent them from being ingested, the anemone being only 1–2 mm in diameter. It is known from a number of localities around the East Anglian and Channel coasts of England (from Lincolnshire to Dorset), and, outside the region, from both North American coasts.

Edwardsia ivelli (Fig. 32)

The only known locality in the world for this small burrowing anemone was the soft mud bed of the Widewater Lagoon in Sussex, but it has not been seen there since 1983 and may therefore be extinct. Nevertheless, since it is only 1 mm wide and up to 20 mm long when fully extended, and since it is rather unlikely that the species evolved in the Widewater Lagoon, it may well be living, unnoticed, in other localities.

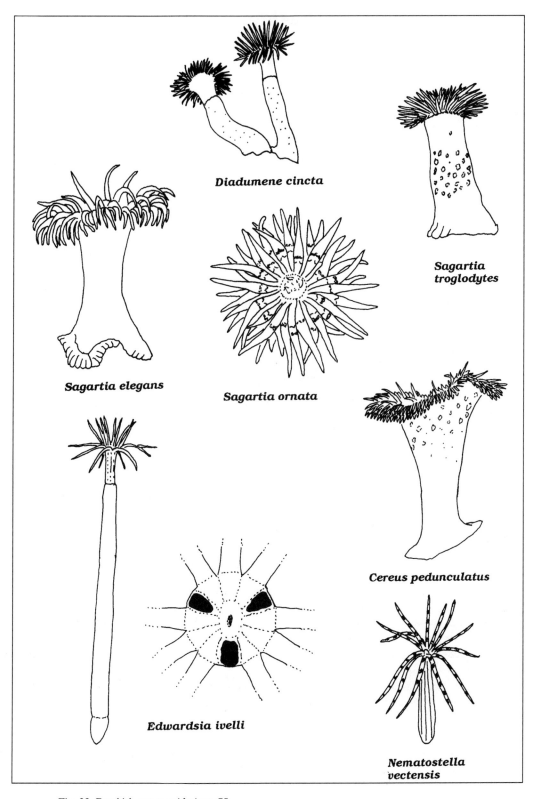

Diadumene cincta

Sagartia troglodytes

Sagartia elegans

Sagartia ornata

Cereus pedunculatus

Edwardsia ivelli

Nematostella vectensis

Fig. 32. Brackish-water cnidarians (V).

Flatworms (planarians) and nemertines (nemerteans or ribbonworms)

Soft, unsegmented worms, flat in section, usually with a mucous covering and often with eyespots, that glide smoothly across surfaces by means of cilia or undulations of the ventral surface. Ball & Reynoldson (1981), Prudhoe (1982) and Gibson (1982) describe British species. See Fig. 33.

1 A Animals of length less than 3 times their breadth (when alive and actively moving); body whitish in colour, delicate, broadest anteriorly .... *Leptoplana tremellaris*
  B Animals of length more than 3 times their breadth (when alive and actively moving) 2

2 A Animals of length less than 25 times their breadth 3
  B Animals of length more than 25 times their breadth 9

3 A Body yellowish with a square black or dark brown patch on the head
  .... *Tetrastemma melanocephalum*
  B Without a square dark patch on the head contrasting with the background colour................ 4

4 A Head with 1 pair of eyes... 5
  B Head with 2 pairs of eyes (in one case, 2 pairs of double eyes).. 7

5 A Eyes surrounded by extensive lightly pigmented patches; head with a pair of short, rounded, anterior 'tentacles', giving it a trilobed appearance .... *Procerodes littoralis*
  B Eyes not surrounded by extensive lightly pigmented patches; head without a pair of tentacles 6

6 A Eyes set back from the anterior margin, in the neck region; body slender, pale coloured; in relatively salty water.. *Uteriporus vulgaris*
  B Eyes close to the anterior margin, within the head; body squat, darkly coloured; in the most dilute of brackish waters .... *Dugesia lugubris/polychroa*

7 A Eyes single, arranged at the corners of a square or rectangle... 8
  B Eyes double, anterior pair closer together than posterior pair .... *Prostomatella obscurum*

8 A Anterior eyes larger than posterior ones; head bilobed; body flattish, pale, without marked colour pattern; occurring under stones at the water line .... *Prosorhochmus claparedii*
  B Eyes equisized; head not bilobed; body firm and oval in section; usually with distinct colour pattern of longitudinal stripe/s, transverse streaks, or series of blotches; occurring below the water line .... *Oerstedia dorsalis*

9 A Eyes arranged in 4 clumps, each clump at the corner of a square on the head (or the 2 anterior clumps extending as a broad band further anteriorly), each clump with several to many eyes; body colour white, grey, pink or orange, not usually more than 40 mm long and 1–2 mm wide .... *Amphiporus lactifloreus*
  B Eyes, if present, not arranged as above; body more than 50 mm long............................. 10

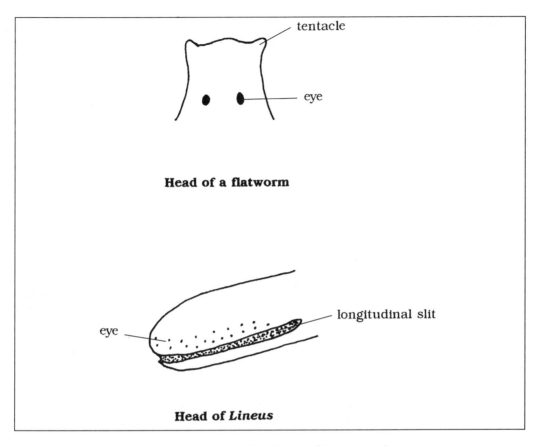

Fig. 33. Anatomical features used in the identification of flatworms and nemertines.

10 A Body thread-like (approx. 0.5 mm wide), pale coloured (white to pale grey); without lateral longitudinal slits on the head 11

B Body dark coloured (reddish brown or olive green through to black); head with longitudinal slits along its sides 12

11 A Head with orange or red pigment at its tip; body up to 6 cm long............. *Cephalothrix rufifrons*

B Head without orange or red pigment; body up to 30 cm long.................... *Cephalothrix linearis*

12 A Body with two narrow longitudinal stripes of white or pale yellow running down the dorsal mid-line .... *Lineus bilineatus*

B Body without such longitudinal stripes 13

13 A Body with more than 10 eyes on each side of the head; body dark brown or black in colour with a flickering purple iridescence... *Lineus longissimus*

B Body with 28 eyes on each side of the head; body reddish or greenish in colour (*Lineus ruber* agg.) 14

14 A When disturbed, body contracts into a tight spiral coil .... *Lineus sanguineus*

B When disturbed, body contracts but does not spirally coil... 15

15 A Body colour reddish .... *Lineus ruber*
B Body colour greenish .... *Lineus viridis*
[These two species differ mainly in the form of their larval development, and are difficult to distinguish morphologically.
L. viridis *has commonly been recorded as* L. gesserensis.*]*

PLATYHELMINTHES
Turbellaria
Tricladida
Procerodidae
Procerodes littoralis (Fig. 34)
A North Atlantic species also known as *Gunda ulvae* that typically occurs in areas subject to rapidly changing salinity, for example wherever fresh water streams or lagoon outlet-discharges empty into the sea, there being found under stones. It also occurs in the Baltic. <10 mm long.

Uteriporidae
Uteriporus vulgaris (Fig. 34)
A common inhabitant of the surface of lagoonal and estuarine sediments, including on salt-marshes, over the southern half of the region. Its salinity range is poorly known but it certainly extends from sea water down to some 10‰. <10 mm long.

Dugesiidae
Dugesia lugubris/polychroa (Fig. 34)
These two species, which have often been confused in the past, are essentially fresh water in habitat, but may penetrate into the more dilute brackish waters, such as lagoons with salinities below some 8‰. Most such brackish-water occurrences are probably of *D. polychroa*. Up to 20 mm long.

Polycladida
Leptoplanidae
Leptoplana tremellaris (Fig. 34)
A widespread species occurring on most types of shores, including mussel beds, from the Arctic to the Mediterranean and Black Seas. Within brackish water, it has been recorded from The Fleet, Dorset, where it consumes gastropod molluscs such as *Hydrobia* and (small) *Littorina* and *Rissostomia*. It presumably occurs elsewhere as well. Up to 25 mm long.

NEMERTEA
Palaeonemertea
Cephalothricidae
Cephalothrix rufifrons (Fig. 34)
Occasionally recorded from the mouths of estuaries, including in *Zostera* beds. Occurs through northwest Europe to the Mediterranean.

Cephalothrix linearis (Fig. 34)
A North Atlantic species that has occasionally been recorded from the mouths of estuaries in sediments ranging from sand to mud.

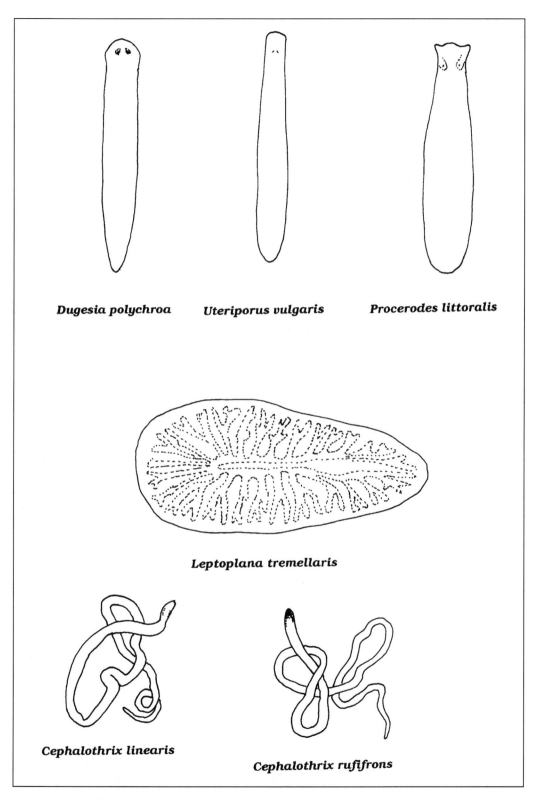

Dugesia polychroa   *Uteriporus vulgaris*   *Procerodes littoralis*

Leptoplana tremellaris

Cephalothrix linearis

Cephalothrix rufifrons

Fig. 34. Brackish-water flatworms and nemertines (I).

Heteronemertea
Lineidae
Lineus ruber (Fig. 35); *Lineus viridis* (Fig. 35); *Lineus sanguineus* (Fig. 35)
These three species have often been confused, and the additional name *L. gesserensis* has been applied to any one of them (especially to *L. viridis*). All three occur in muddy sediments in estuaries and lagoon-like brackish waters, *L. ruber* in particular extending down to some 8‰ salinity, and *L. sanguineus* perhaps being the least euryhaline. They are presumably predatory but their brackish-water diets are largely unrecorded (small gastropods have been observed in their guts, and crustaceans have also been suggested). Densities of *L. ruber* can attain almost 2000 individuals/m². Usually <20 cm long.

Lineus longissimus (Fig. 35)
A widely distributed species that occurs throughout northwest Europe including in the Baltic, but which otherwise has not often been recorded from brackish water. It is usually to be found under stones on mud or muddy sand. *L. longissimus* is notable for being the longest animal known: lengths of 10 m are common and one "not half uncoiled" specimen was measured to be 27 m.

Lineus bilineatus (Fig. 35)
Occurs in sediments ranging from gravel to muddy sand, and occasionally just extends into the >22‰ salinity regions of estuaries. Up to 50 cm long.

Hoplonemertea
Amphiporidae
Amphiporus lactifloreus (Fig. 35)
An Arctic to Mediterranean species that penetrates the Baltic as far as Kiel Bay (where it forms part of the *Zostera* fauna) and that has been recorded from several estuaries and from sandy or gravelly lagoons. Outside the Baltic, some 20‰ appears to be its lower limit. <10 cm long.

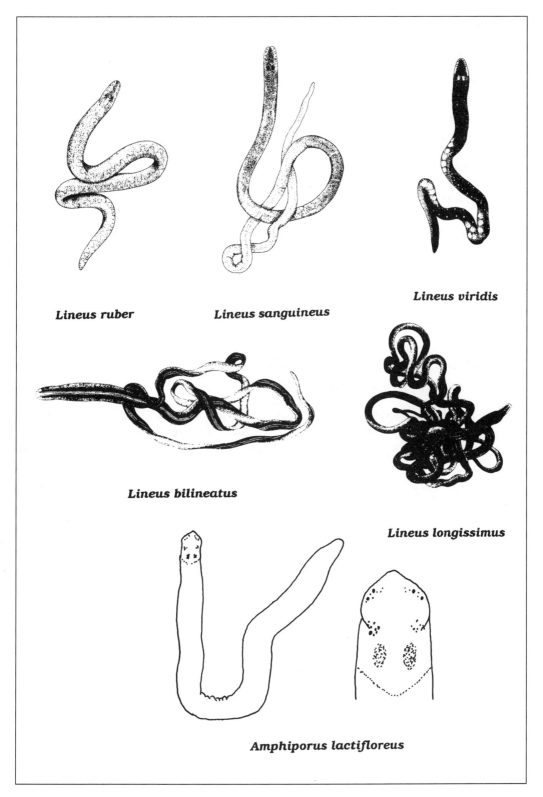

Lineus ruber

Lineus sanguineus

Lineus viridis

Lineus bilineatus

Lineus longissimus

Amphiporus lactifloreus

Fig. 35. Brackish-water nemertines (II).

Prosorhochmidae
Oerstedia dorsalis (Fig. 36)
Mainly a marine rock-pool or kelp-holdfast species that has been recorded from beneath a log in a brackish-water lagoon. <30 mm long.

Prosorhochmus claparedii (Fig. 36)
A southwestern species known from brackish water only as part of the waterline shingle fauna in The Fleet, Dorset. <40 mm long.

Tetrastemmatidae
Tetrastemma melanocephalum (Fig. 36)
A widely distributed species that in northwest European brackish waters occurs in the mouths of estuaries, in sediments ranging from gravel to mud and amongst *Zostera*. Densities of up to 500 individuals/m^2 and a diet of *Corophium* have been recorded. Up to 60 mm long.

Prostomatella obscurum (Fig. 36)
This species is confined to Danish waters in the east of the region, its distribution being centred on the Baltic in which it occurs in salinities of down to 6‰. It is viviparous. *c.* 10 mm long.

Nematodes (roundworms)

Very small, unsegmented worms, round in section and with more or less pointed ends (Fig. 36), that move via somewhat stiff C- or S-shaped thrashing. Very few (only the largest) species, e.g. *Enoplus brevis* at 7 mm long but only 150 μm across, are likely to be retained by a 500 μm mesh sieve, and these are translucent. Keys are not yet available to all the free-living species, but Platt & Warwick (1983; 1988) describe many of them.

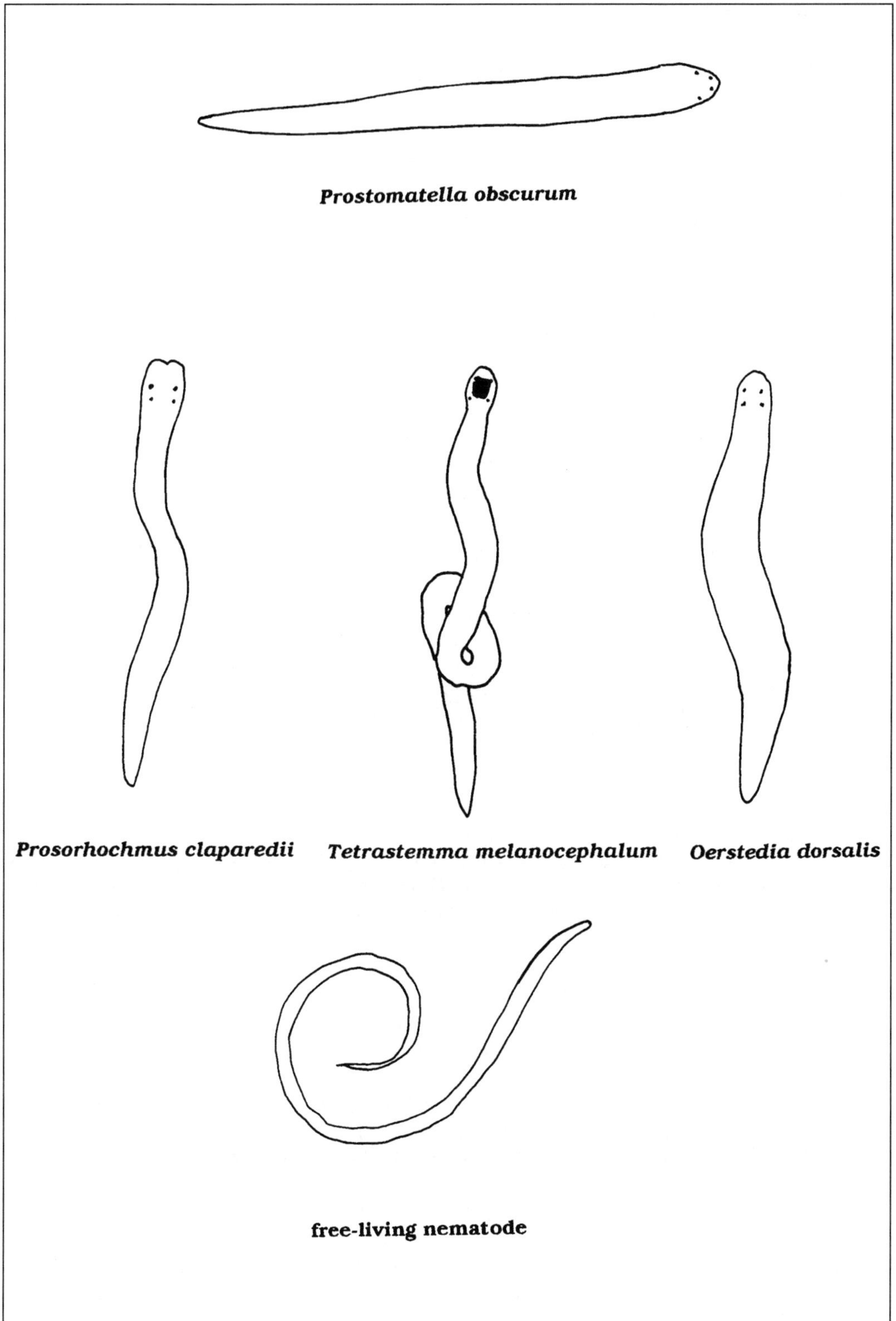

Prostomatella obscurum

Prosorhochmus claparedii   *Tetrastemma melanocephalum*   *Oerstedia dorsalis*

free-living nematode

Fig. 36. Brackish-water nemertines (III) and nematodes.

Annelids

Polychaetes (bristleworms)

A large assemblage of diverse segmented worms of two broad types: (a) highly mobile species with short, sensory tentacles and often eyes on the head, jaws that can be everted through the mouth, and a relatively uniform body, the sides of which bear pairs of paddle- or limb-like projections (parapodia); and (b) less mobile species with the anterior and posterior parts of their bodies of different form, at least the anterior part bearing reduced parapodia and sometimes gills, and with often long, feeding tentacles or palps at their anterior end. Some species, however, are superficially similar to earthworms and can be distinguished from that group only by the absence of a clitellum and by the presence of reduced parapodia (in the form of ridges running down the side of some of the segments). Up to 30 cm in length (exceptionally longer) and especially associated with the sediments, including occurring in tubes constructed of sand grains, of jelly and of other materials. See Fig. 37.

n.b. as elsewhere, this key is designed for use with living animals: the palps referred to in couplet 23 et seq., the gills of spionids, ampharetids and others, and the scales of Harmothoë, *amongst other appendages, are likely to be shed if their owner is dropped into, for example, alcohol or other preservative. The key should therefore not be used with dead or preserved material.*

1 A Animals living within a white, calcareous tube 2

   B Animals tube-dwelling or free-living, but not inhabiting a calcareous tube 3

2 A Tube tightly coiled .... spirorbids

   B Tube not coiled .... *Ficopotamus enigmatica*

3 A Animal inhabiting a large tube made from cemented sand grains, of which the topmost few centimetres project above the sediment surface and of which the mouth forms a ragged tassel, again of sand grains; remainder of tube buried in sediment; animal up to 30 cm long .... *Lanice conchilega*

   B Animal free-living or tube-dwelling; if tube-dwelling, then tube not as above 4

4 A Animal inhabiting one of many contiguous, ± straight tubes made of cemented sand grains that are attached to hard substrata and form reefs, the openings of the tubes forming a honeycomb pattern; animal up to 40 mm long .... *Sabellaria alveolata*
       [The related S. spinulosa occurs sublittorally, forming 'ross' on stones and shells. Its disorderly tubes are rarely straight and often have the openings irregularly arranged.]

   B Animal free-living or tube-dwelling; if tube-dwelling, then tube not as above 5

5 A Large, oval worm (some 3× as long as broad) with the dorsal surface covered by dense, fur-like, grey-brown chaetae; up to 20 cm long and 7 cm broad .... *Aphrodita aculeata*

   B Elongate worms with lengths >4× as long as broad and with their dorsal surface obscured by scales or visible, but not covered by felt of chaetae 6

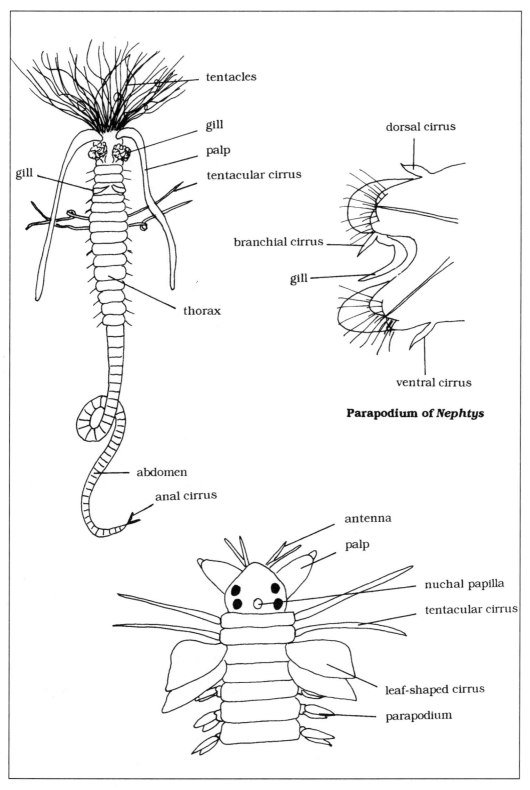

Fig. 37. Anatomical features used in the identification of polychaetes.

6 A Dorsal surface of body concealed beneath overlapping scales 7
  B Dorsal surface of body not concealed beneath scales 12

7 A With 12 pairs of dorsal scales .... *Lepidonotus squamatus*
  B With 15 or more pairs of scales 8

8 A With >40 pairs of dorsal scales .... *Pholoë minuta*
  B With 15 pairs of scales 9

9 A Without scales on the last 4 segments; commensal in burrows of terebellid polychaetes
    .... *Gattyana cirrhosa*
  B Whole dorsal surface obscured by scales.. 10

10 A 1st pair of scales almost white (often with central black spot), contrasting with those following; up to 15 mm long... *Harmothoë spinifera*
  B 1st pair of scales not white and not contrasting with those following; up to 70 mm long...... 11

11 A Each segment with 1 long and 2 short transverse bands of pigment; up to 70 mm long; often commensal with *Arenicola*; east of region only .... *Harmothoë (Antinoëlla) sarsi*
  B Without segmental bands of pigment; up to 50 mm long; usually in association with terebellid polychaetes .... *Harmothoë imbricata*

12 A Without antennae, palps or tentacles on the anteriormost 3 apparent segments 13
  B Anteriormost 3 apparent segments with antennae, palps or tentacles (the antennae of some species may be visible only with the microscope) 21

13 A Body minute (up to 6 mm long, <1 mm wide), with 4 regions: (i) a 2-segmented head, (ii) a collar with conspicuously polygonal surface pattern, both (i) and (ii) without appendages; (iii) a 6-segmented 'thorax', each segment with a pair of dorsal tentacles, decreasing in size posteriorly; and (iv) a 30-segmented, appendage-less 'abdomen'........ *Psammodrilus balanoglossoides*
  B Body not as above 14

14 A Body terminating anteriorly in a median sensory papilla .... *Paraonis fulgens*
  B Body without a terminal sensory papilla 15

15 A Body divided into 2 regions, a fatter anterior 'trunk' portion and a thin tail region, the posterior part of the trunk with 13 pairs of tufted gills; all but the first few trunk segments divided into 5 annuli (the lugworm)... *Arenicola marina*
  B Gills, if present, not as above; segments not divided into 5 annuli 16

16 A Body long, thin, earthworm-like, without any projecting gills; at least the anterior portion red 17
  B With projecting finger-like gills somewhere along body; body sometimes orange-pink but not red 18

17 A With an anterior region of 9 short, chaeta-bearing segments, followed by a posterior region of 80+ more elongate segments without evident chaetae; body blood-red .... *Capitella capitata*

ANNELIDS 81

B With an anterior region of 12 short, chaeta-bearing segments, followed by a posterior region of >100 more elongate segments without evident chaetae; dull red anteriorly, greenish or yellowish posteriorly .... *Heteromastus filiformis*

18 A Long orange-pink body without a ventral gutter; with gills from segment 13 (± 4) to end of body; with 200+ segments .... *Scoloplos armiger*
   B Short worms with a ventral gutter along the body, at least posteriorly 19

19 A Ventral gutter extends along whole of the body; with 26 or 27 chaeta-bearing segments; 3 eyes on the head; very small, <8 mm long .... *Armandia cirrhosa*
   B Ventral gutter present only on posterior part of the body 20

20 A With <26 chaeta-bearing segments, <11 pairs of gills and 1 anal papilla .... *Ophelia rathkei*
   B With 32 chaeta-bearing segments, 15 pairs of gills and 2 anal papillae .... *Ophelia bicornis*

21 A Anteriormost segment with 4 very small antennae, without other appendages at the head end.. 22
   B Anterior appendages not as above 23

22 A Head conical, annulated, with 4 antennae in a cross at its tip; body cylindrical <80 mm long .... *Glycera capitata*
   B Head rectangular, with 4 antennae, one at each anterior corner; body flattened, silvery, up to 25 cm long .... *Nephtys* spp. (see separate key below)

23 A Head end with single pair of long and sometimes coiled, grooved palps (much longer or stouter than any other appendages present anteriorly) 24
   B Appendages of head end not dominated by a pair of palps 38

24 A Head region anterior to the palps, flat and broad (wider than rest of body), without eyes; palps inserted ventrally, with row of papillae down one side .... *Magelona mirabilis*
   B Head not broader than body 25

25 A Other tentacle-like appendages, besides palps, present anteriorly; 'tentacles' present on most body segments.. 26
   B Palps form the only tentacle-like appendages at head end of body 27

26 A Without eyes; with 200+ segments; palps on 1st chaeta-bearing segment .... *Tharyx marioni*
   B With pair of lateral eyes; with <150 segments; palps on segment in front of 1st chaeta-bearing one .... *Caulleriella zetlandica*

27 A Animal bores into calcareous rock and shells .... *Polydora ciliata*
   B Animal living in sediment.. 28

28 A 5th chaeta-bearing segment not different from its neighbours 29
   B 5th chaeta-bearing segment markedly different from all others, e.g. lacking gills and lateral lobes, but with very large chaetal spines 36

29 A Without finger-like gills arching over dorsal surface .... *Spiophanes bombyx*
   B With finger-like gills arching over dorsal surface of at least one and often most segments 30

30 A Gills present only on 1st chaeta-bearing segment .... *Streblospio shrubsoli*
   B Gills present on more than one segment 31

31 A Gills either in form of 7–8 pairs beginning on chaeta-bearing segment 11–20 (females) or 20–28 pairs in the middle third of the body plus 1 pair on 2nd chaeta-bearing segment (males); with 4 stout terminal processes; inhabits long tube of fine sand grains; up to 15 mm long .... *Pygospio elegans*
   B Gills from 1st or 2nd chaeta-bearing segment down at least first third of body 32

32 A With terminal funnel-like membrane around the anus, without anal cirri 33
   B With 4–8 terminal petal-like or cylindrical cirri around the anus 34

33 A Gills present along length of body; head without median tentacle; up to 80 mm long
   .... *Scolelepis squamata*
   B Gills absent from last third of body; head with median tentacle; up to 160 mm long
   .... *Scolelepis foliosa*

34 A Gills present only on anterior part of body .... *Marenzelleria viridis*
   B Gills present along whole length of the body 35

35 A With 4 anal cirri; head without anterior frontal horns; up to 30 mm long .... *Spio filicornis*
   [Sometimes recorded under the same S. martinensis.*]*
   B With 6–8 anal cirri; head with 1 pair of anterior frontal horns; up to 60 mm long
   .... *Malacoceros fuliginosus*

36 A Chaetal spines of 5th chaeta-bearing segment arranged in 2 horseshoe-shaped rows
   .... *Polydora (Pseudopolydora) pulchra*
   B Chaetal spines of 5th chaeta-bearing segment not arranged in horseshoe-shaped rows 37

37 A Gills start on 7th chaeta-bearing segment; with terminal funnel-shaped membrane around the anus .... *Polydora ligni*
   B Gills start on 2nd chaeta-bearing segment; with 2 anal cirri .... *Polydora (Boccardia) ligerica*
   [Polydora (Boccardia) redeki is now considered to be a synonym.]

38 A Head without appendages, but either 1st or one of 4th–7th chaeta-bearing segments with numerous tentacles arising from its dorsal surface; long tentacle-like gills issue from the sides of most segments 39
   B Body not as above 40

39 A Both tentacles and tentacle-like gills arise on 1st chaeta-bearing segment; 4-8 eyes in diagonal row on each side of head; up to 120 mm long .... *Cirratulus cirratus*
   B Tentacles arise from 4th, 5th, 6th or 7th chaeta-bearing segment; tentacle-like gills begin on 1st chaeta-bearing segment; without eyes; up to 200 mm long .... *Cirriformia tentaculata*

40 **A** Elongate mobile worms, with parapodia issuing from sides of segments; anterior segments with eyes, antennae, 2–4 pairs of lateral tentacular cirri and sometimes palps; body not differentiated into regions 41

B Relatively broad-bodied sedentary worms, without parapodia; anterior segments with various tentacles and often gills, without antennae and often without eyes; body divided into at least 2 distinct regions 51

41 **A** Head with 3 antennae, next segment with 2 pairs of lateral tentacular cirri; parapodia each give rise to an elongate, tentacle-like cirrus dorsally; <5 mm long 42

B Head with 2, 4 or 5 antennae, next segment/s with 2–4 pairs of lateral tentacular cirri; parapodia each with finger- or leaf-like dorsal cirrus 43

42 **A** With pair of minute, finger-like palps; median antenna inserted between the 2 anterior eyes; with 3 anal cirri .... *Streptosyllis websteri*

B Palps large and fused together into a triangular mass; antennae all inserted anterior to the eyes; with 2 anal cirri .... *Exogone naidina*
[*Sometimes recorded under the name* E. gemmifera]

43 **A** Head region with 2 antennae, 2 palps, 4 eyes and 4 pairs of tentacular cirri; parapodia with a finger-like, not leaf-like, dorsal cirrus 44

B Head region with 4 or 5 antennae, no palps, 2 eyes, and 2–4 pairs of short to long tentacular cirri; parapodia with a leaf-like dorsal cirrus 47

44 **A** Dorsal lobe of parapodia markedly larger than ventral lobe, at least on posterior segments 45

B Dorsal lobe of parapodia ± same size as ventral lobe 46

45 **A** Uppermost lobe of parapodia increasing in size, and the dorsal cirrus increasingly moving toward the tip of the lobe, on posterior progression; brownish; up to 120 mm long .... *Nereis (Neanthes) succinea*

B Uppermost lobe of parapodia uniformly large along the body, its dorsal cirrus always basal; greenish; up to 300 mm long or more .... *Nereis (Neanthes) virens*

46 **A** Dorsal cirrus elongate, extending beyond tip of parapodia; body solid, dorsal surface rounded; with 2 dorsal eyebrow-like chitinous bars on the everted proboscis; bronze hued; up to 250 mm long .... *Perinereis cultrifera*

B Dorsal cirrus not extending beyond tip of parapodia; body limp with flat dorsal surface; without dorsal chitinous bar on the everted proboscis; colour variable but not bronzy; up to 120 mm long.. *Nereis (Hediste) diversicolor*

47 **A** With 2 pairs of short, conical tentacular cirri, on first segment behind the head 48
B With 4 pairs of long tentacular cirri, 1 pair on 1st, 2 on 2nd and 1 on 3rd segment behind head 49

48 **A** With indistinct nuchal papilla; with red-brown pigment dorsally on a yellowish-white background .... *Eteone picta*

B Nuchal papilla large and obvious; translucent yellowish-white without red-brown patches .... *Eteone longa*

49 A With 4 anterior and 1 median antennae .... *Eulalia viridis*
   B With 4 anterior antennae but without median one... 50

50 A With dark brown and blue transverse bands across every segment dorsally.. *Phyllodoce groenlandica*
   B With conspicuous dark brown spots on dorsal surface and on the leaf-like dorsal cirri
   .... *Phyllodoce maculata*

51 A Very small worms, < 6mm long, with only 10–12 chaeta-bearing segments; head with 1
   pair of tentacle-like palps and 3–4 tentacles on each side, which may bear secondary pinnules, and with 1 pair eyes 52
   B Generally larger worms with >25 chaeta-bearing segments 54

52 A With 4 short pinnule-less tentacles united at their base on each side of the head; without
   pair of eyes on anal segment; up to 6 mm long .... *Manayunkia aestuarina*
   B With 3 tentacles, each with some 6 pairs of pinnules, on each side of the head; with pair of
   eyes on anal segment; up to 3 mm long.. 53

53 A Palps short, partially obscured by triangular ventral lobe .... *Fabricia stellata*
   [Often recorded as Fabricia sabella*]*
   B Palps long, as long as the tentacles, not obscured by ventral lobe .... *Fabriciola baltica*

54 A With circular funnel or crown of 80+ stiff tentacles 55
   B Worms with their tentacles not arranged in a stiff circle... 56

55 A Inhabiting a thick, transparent tube of mucus .... *Myxicola infundibulum*
   B Inhabiting a tube of cemented mud particles .... *Sabella pavonina*

56 A Head with numerous, very long, mobile tentacles that roam over or in the substratum; tentacles contractile but not capable of being withdrawn into the body; with 1 or 3 pairs of
   branched gills immediately posterior to the tentacles 57
   B Head with short tentacles that appear to issue from (and can be withdrawn into) the
   mouth and with 3–4 pairs of simple, elongate, finger-like gills 58

57 A With 1 pair of gills, each gill comprising 4 branches with kidney-shaped lamellae issuing
   from a common, fat, cylindrical stalk .... *Terebellides stroemi*
   B With 3 pairs of bush-like gills .... *Amphitrite figulus*

58 A Very small, <5 mm long; with 3 pairs of gills, 16 thoracic and 13–19 abdominal chaeta-
   bearing segments .... *Alkmaria romijni*
   B Large worms, with 4 pairs of gills.. 59

59 A With long abdominal region (60 chaeta-bearing segments), without anal cirri; gills united by
   web for 50% of their height; without bundle of golden chaetae in front of each group of
   gills.. *Melinna palmata*
   B With short abdominal region (12 chaeta-bearing segments), with several cirri around anus;
   gills not united by a web; with a bundle of golden chaetae in front of each group of gills
   .... *Ampharete grubei* [Often recorded as *Ampharete acutifrons*]

Key to *Nephtys* species

(The 4 Nephtys species that may be found in brackish waters are far from easy to distinguish when alive. With care, however, they can be separated on the form of their parapodia (see Fig. 40), as below)

1 **A** 1st chaeta-bearing segment with a dorsal cirrus of same size as ventral cirrus...................... 2
  B 1st chaeta-bearing segment without a dorsal cirrus.. 3

2 **A** Ventral cirrus as large as the gill; up to 150 mm long............................... *Nephtys longosetosa*
  B Ventral cirrus shorter than the gill; up to 250 mm long .... *Nephtys caeca*

3 **A** Branchial cirrus as long as the gill on posterior parapodia; up to 100 mm long.... *Nephtys cirrosa*
  B Branchial cirrus always much shorter than the gill; up to 200 mm long............ *Nephtys hombergi*

POLYCHAETA
Phyllodocida
Aphroditidae

Aphrodita aculeata (sea mouse) (Fig. 38)

The sea mouse occasionally penetrates down to salinities of some 18‰ in the estuaries of north-western Europe, although it is never common. The generic name is often spelt with a terminal 'e'.

Polynoidae (scaleworms)

Harmothoe spinifera (Fig. 38)

An uncommon scaleworm in estuaries, occasionally recorded from waters of >22‰.

Harmothoe (Antinoella) sarsi (Fig. 38)

Recorded from the southern shore of the North Sea from The Netherlands, Germany and Denmark, in estuarine salinities of down to 18‰; but mainly an Arctic and Baltic species, where it is known in waters of down to 4‰. In the Netherlands, it is particularly associated with the burrows of *Arenicola marina* although elsewhere it occurs free, on and in the sediment.

Harmothoe imbricata (Fig. 38)

A widespread scaleworm which, in estuaries, inhabits mussel beds and the massed tubes of *Lanice*, amongst other microhabitats. Regularly inhabits salinities of down to 16‰ and more rarely of down to 5‰. Length <65 mm.

Lepidonotus squamatus (Fig. 38)

A widely distributed scaleworm, which in northwest European estuaries is especially associated with mussel beds, although it also occurs under stones and amongst *Sabellaria*, in waters of down to 18‰, and more locally 12‰. Length <50 mm.

Gattyana cirrhosa (Fig. 38)

Present in the mouths of estuaries, in waters of above 22‰, throughout the region wherever its terebellid hosts (particularly *Amphitrite*) occur. In the mouth of the Baltic it can tolerate 12–15‰. Length <50 mm.

Sigalionidae

Pholoe minuta (Fig. 38)

Occurs in shelly estuarine muds throughout the region down to some 15‰ salinity (and locally less). Length <25 mm.

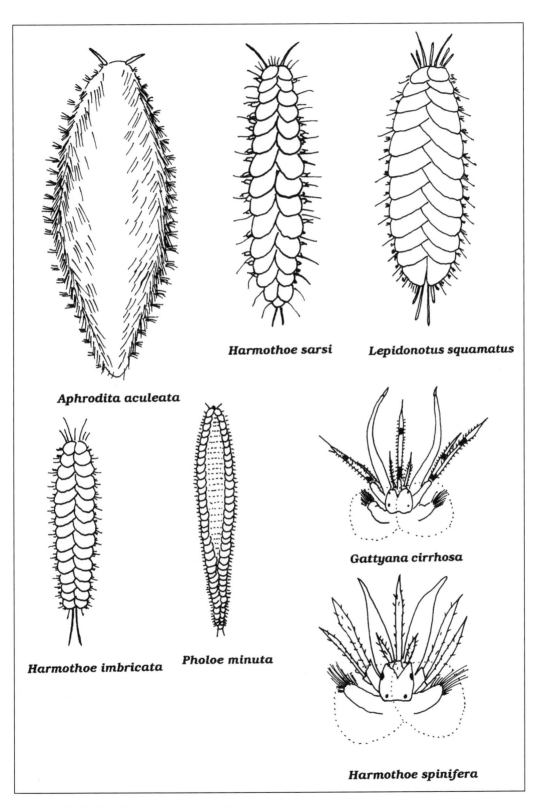

Fig. 38. Brackish-water polychaetes (I).

Phyllodocidae

Phyllodoce groenlandica (Fig. 39)

This circum-Arctic species, which is often placed in the genus *Anaitides*, occurs throughout the region in relatively sandy muds, in estuarine waters usually of above 22‰. May achieve 0.5 m length.

Phyllodoce maculata (Fig. 39)

Similarly to *P. groenlandica*, this Arctic species, which is often placed in the genus *Anaitides*, occurs throughout the region in relatively sandy muds, in estuarine waters usually of above 22‰. It also occurs in gravel and in the more saline mussel beds. Length <100 mm.

Eteone longa (Fig. 39)

E. longa is a common predator of spionids in estuarine muds and muddy sands throughout the region, and more rarely occurs in non-tidal lagoons. It occurs down to 18‰, and locally to 8‰. Length up to 150 mm.

Eteone picta (Fig. 39)

More local than *E. longa* above, this western species (also placed in the genus *Mysta*) occurs from the Channel westwards in gravelly sediments and it too probably consumes spionids. Length <60 mm.

Eulalia viridis (Fig. 39)

Although best known as a rocky-shore species, *E. viridis* inhabits estuarine mussel beds in salinities of down to 18‰. Length <150 mm.

Glyceridae

Glycera capitata (bloodworm) (Fig. 39)

This bloodworm is recorded from all the world's oceans, and occurs in the mouths of northwest European estuaries in waters of more than 22‰ salinity. Like its relatives, it is an active carnivore which swims with spiral body movements and often coils up when disturbed.

Syllidae

Exogone naidina (Fig. 39) and *Streptosyllis websteri* (Fig. 39)

Somewhat unusually for syllids, these two species occur commonly on and in soft sediments in estuarine waters of down to some 18‰. *E. naidina* is also frequently referred to as *E. gemmifera*. Length up to 5 mm.

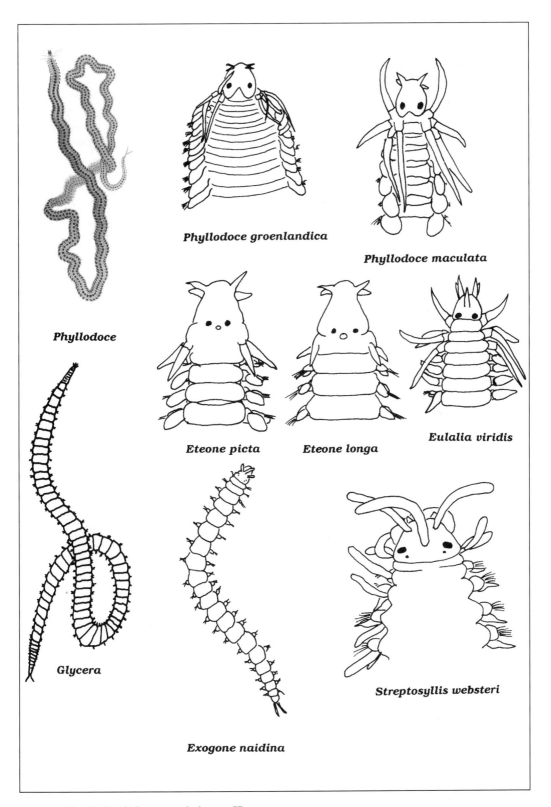

Phyllodoce groenlandica

Phyllodoce maculata

Phyllodoce

Eteone picta

Eteone longa

Eulalia viridis

Glycera

Streptosyllis websteri

Exogone naidina

Fig. 39. Brackish-water polychaetes (II).

Nereidae

Nereis diversicolor (ragworm) (Fig. 40)

Perhaps *the* northwest European brackish-water polychaete, being widespread in soft sediments in tidal and non-tidal habitats throughout the region in waters of down to 5‰ or less salinity. It is an opportunistic predator/herbivore/scavenger and hunter/deposit-feeder/suspension-feeder/browser that spends its larval life in the surface layers of the sediments, and when larger constructs large, branched burrow systems with multiple openings to the surface. Densities may achieve 100 000 individuals/m^2. Like many other successful estuarine invertebrates it is semelparous. Where *Nereis* is split into several genera, this species is placed in *Hediste* .

Nereis virens (king rag) (Fig. 40)

Nereis (or *Neanthes*) *virens* is a much larger worm than the above species (>3× longer) which, when adult, can deliver a painful bite. Although biologically similar to *N. diversicolor*, it is more of a specialist carnivore and inhabits the more saline halves of estuaries (salinities of more than 17‰).

Nereis succinea (Fig. 40)

Nereis (or *Neanthes*) *succinea* is an Atlantic species that is in several respects intermediate between *N. diversicolor* and *N. virens*, and may interact with them. For this reason it is somewhat sporadic in occurrence (both spatially and temporarily), although it can occur over a wide salinity range, down to 5‰ locally, and in all sediment types including in mussel beds.

Perinereis cultrifera (ragworm) (Fig. 40)

This widespread species tends to replace *N. diversicolor* in more saline environments (salinities >20‰), and may do so in the mouths of estuaries and in *Zostera* meadows, especially in regions outside the North Sea.

Nephtyidae (catworms)

Nephtys hombergi (Fig. 40); *Nephtys caeca* (Fig. 40); *Nephtys cirrosa* (Fig. 40); *Nephtys longosetosa* (Fig. 40)

Nephtys species penetrate into the mouths of estuaries and estuarine lagoons until the salinity falls to some 20‰, with *N. hombergi* and *N. caeca* occasionally extending into waters of below 18‰. They inhabit sediments of muddy sand, where they are probably generalist predators/scavengers/deposit-feeders, and may locally displace *N. diversicolor*.

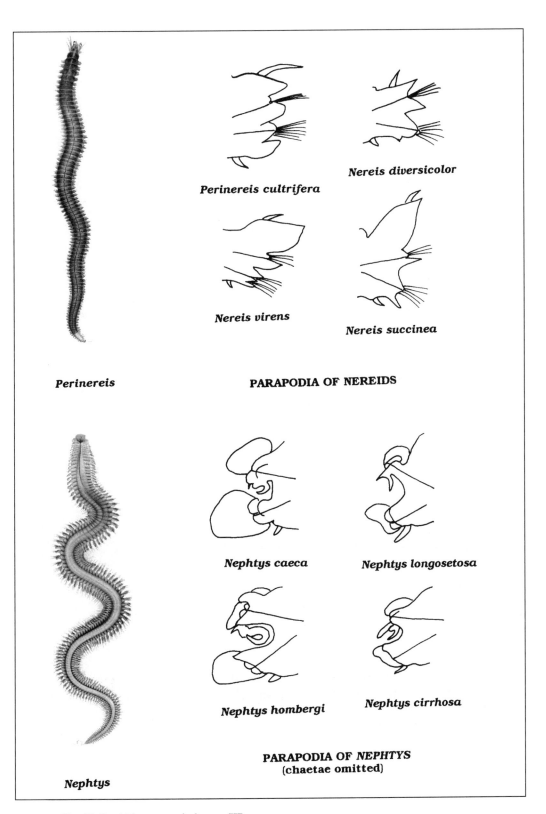

Fig. 40. Brackish-water polychaetes (III).

Orbiniida
Orbiniidae
Scoloplos armiger (Fig. 41)

A common inhabitant of muddy sands and regions of submerged vegetation in both estuaries and lagoon-like habitats in waters of down to 15‰, and locally to less than 10‰. Besides northwest Europe, it is found in all oceans of the world. Length <125 mm.

Paraonidae
Paraonis fulgens (Fig. 41)

An inhabitant of clean sands usually in waters of above 22‰, it is therefore a relatively uncommon species in brackish waters. *P. fulgens* constructs burrow systems that include characteristic and distinctive spiral meanders in the bedding planes of the sand. Length <30 mm.

Spionida
Spionidae
Streblospio shrubsoli (Fig. 41)

A small, characteristically brackish-water species that inhabits low salinities down to 4‰ in both estuaries and lagoon-like habitats, in which it occurs in mud tubes attached to *Ruppia* and in soft sediments. It can be instantly recognised by its two pairs of long anterior 'tentacles'. Length <10 mm.

Malacoceros fuliginosus (Fig. 41)

A northeast Atlantic and Mediterranean species that burrows in thick, black muds, *Zostera* debris and in muddy sands. In spite of its preference for this typically estuarine type of habitat, it does not seem to be widespread in brackish waters, and most of the records of it are from sites with salinities >20‰, including from Channel coast lagoons.

Scolelepis foliosa (Fig. 41)

Occurs in relatively clean sandy sediments and, as such, is found only at the mouths of estuaries, down to salinities of some 23‰.

Scolelepis squamata (Fig. 41)

Like *S. foliosa* above, an inhabitant of relatively clean sands and hence found only near the mouths of some estuaries, in salinities of down to some 18‰.

Spiophanes bombyx (Fig. 41)

Like the *Scolelepis* species, it occurs in sands, but *Spiophanes* is able to tolerate relatively muddy sands. Nevertheless, it appears to be restricted to high-salinity areas: >22‰. Length <60 mm.

Marenzelleria viridis (Fig. 41)

A species native to the eastern coast of North America that has recently been reported from several North Sea estuaries (the Forth, Tay, Ems, Weser and Elbe), mostly from areas of 4–16‰. It inhabits vertical, mucus-lined burrows in soft sediments. Early records are under the generic name *Scolecolepides*. Length up to 100 mm.

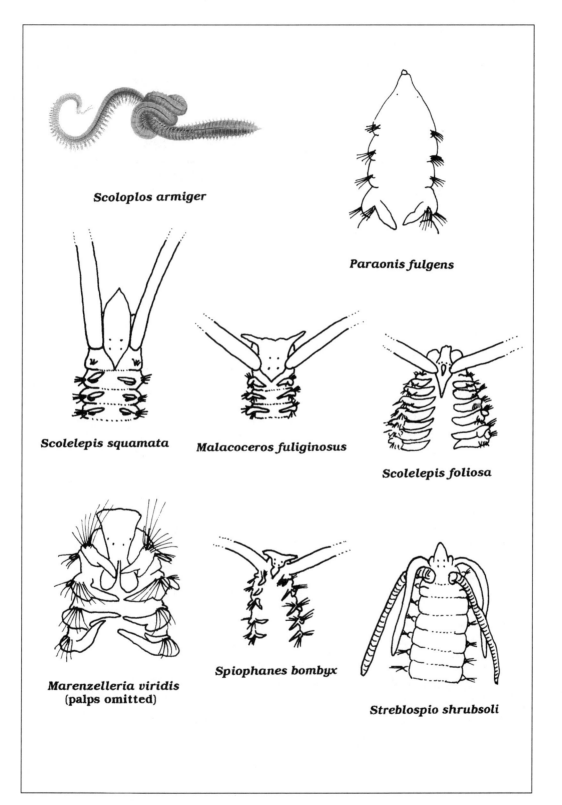

Fig. 41. Brackish-water polychaetes (IV).

Spio filicornis (Fig. 42)

S. filicornis (including *S. martinensis* within this species) is known from the Arctic to the Mediterranean, in waters of down to some 18‰ (and lower in the Baltic). Like many other spionids, however, it is restricted to sands and muddy sands, and hence is not a common or abundant member of the brackish-water fauna.

Polydora (Pseudopolydora) pulchra (Fig. 42)

An inhabitant of the mouths of estuaries (down to salinities of some 22‰ and locally <18‰) and muddy or sandy mud marine areas through the North Atlantic and Mediterranean. In northwest Europe it is recorded from the southern and southeastern shores of the North Sea (Netherlands, Germany and Denmark) in tubes of mud and mucus in the sediment. Length <10 mm.

Polydora (Boccardia) ligerica (Fig. 42)

Inhabits mucus-and-mud tubes in muddy sediments in both tidal and non-tidal brackish waters down to salinities of 0.5‰, and rarely recorded in salinities of over 18‰. Distributed from the Loire Estuary, France, through the Low Countries, to Germany and Denmark although apparently only once from the British Isles (Severn Estuary). *P. ligerica* described from the Loire is now thought to be the same species as *P. redeki* (described from the inland Alkmaarder Meer in the Netherlands). Length up to 30 mm.

Polydora (Polydora) ligni (Fig. 42)

Very similar to *P. ciliata* and often confused with it, the two species differing mainly in habitat and in extent of penetration of brackish water. *P. ligni* constructs tubes of mud and mucus in brackish-water sediments, on *Zostera, Ruppia* and on algae, in salinities of down to 2‰. It occurs in both the North Atlantic and North Pacific, but in northwestern Europe is so far known mainly from the southern and southeastern shores of the North Sea (Netherlands, Germany and Denmark) in both tidal and tideless lagoonal waters. It also occurs in the Severn and Tamar Estuaries, and is probably much more widespread that the current records indicate. Length up to 30 mm.

Polydora (Polydora) ciliata (Fig. 42)

Morphologically very similar to *P. ligni* (see above). *P. ciliata*, however, inhabits tubes in mollusc shells and other calcareous material, in salinities of down to 18‰. Occurs throughout the region. Length up to 30 mm.

Pygospio elegans (Fig. 42)

A characteristic and widespread inhabitant of the sandier brackish-water sediments, in which it constructs a fine, cohesive tube of sand grains. It may be found in both estuarine and lagoonal types of habitat in waters of down to some 4‰ salinity and, for short periods, of down to 2‰, and in densities of up to some 40 000 individuals/m^2. It probably captures food particles by both suspension- and deposit-feeding.

Magelonidae

Magelona mirabilis (Fig. 42)

Like the larger spionids, which it resembles, *Magelona* is an inhabitant of relatively clean sands, and hence only occurs near the mouths of estuaries, in waters of >22‰ salinity. Length <175 mm.

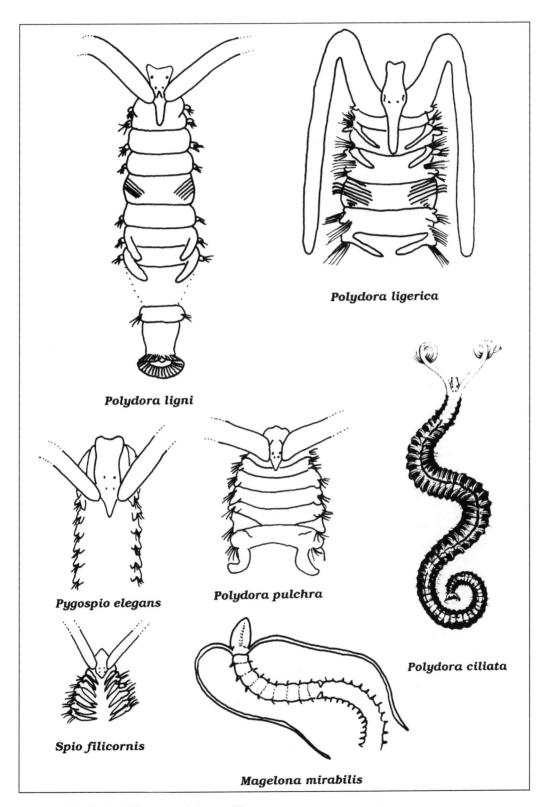

Fig. 42. Brackish-water polychaetes (V).

Cirratulidae

Tharyx marioni (Fig. 43)

Penetrates further into brackish waters than most cirratulids, being found down to 12‰, and lower for short periods. Occurs widely in mud both in estuaries and in non-tidal habitats. Length <100 mm.

Cirratulus cirratus (Fig. 43)

Widespread in mud or muddy sand in sheltered habitats, including lagoons, but not penetrating estuaries to any great extent (absent from below some 20‰).

Cirriformia tentaculata (Fig. 43)

Like *Cirratulus,* widespread in mud, muddy sand or gravelly mud, and in *Zostera* meadows in sheltered habitats, including lagoons, but only found near the mouths of estuaries. Known under the name *Audouinia* in the older literature.

Caulleriella zetlandica (Fig. 43)

A little-known species found in clean sands near the mouths of estuaries, and elsewhere, in the west of the region; there is one record from an English North Sea coast lagoon. Length <50 mm.

Psammodrilida

Psammodrilidae

Psammodrilus balanoglossoides (Fig. 43)

Lives in a mucus-and-detritus tube in sandy sediments, and known from the Baltic in salinities of down to 12‰; elsewhere, however, seems to occur only in salinities of >22‰ and therefore in the mouths of estuaries.

Capitellida

Capitellidae

Capitella capitata (Fig. 43)

A widespread inhabitant of muds and muddy sands that is especially abundant in areas of organic pollution. Even in non-polluted areas, densities can exceed 250 000 individuals/m^2. It occurs in the seawards halves of estuaries (salinities >18‰) and under corresponding conditions in non-tidal brackish waters. Length up to 120 mm.

Heteromastus filiformis (Fig. 43)

A northeast Atlantic and Mediterranean species that occurs in estuarine muddy sands down to salinities as low as 5‰, although it is relatively rare below 18‰. Length up to 180 mm.

Arenicolidae

Arenicola marina (lug- or lobworm) (Fig. 43)

Abundant in estuarine and non-tidal sediments of muddy sand down to salinities of some 18‰ (and below, to 8‰ in the Baltic and some other localities); the depressions and mounds of its burrow systems structure the substratum of many lagoons. It consumes the sand at the head end of its burrow, but by maintaining a water current through the head shaft it not only permits the sand there to filter fine particles but also creates optimum conditions for the growth and multiplication of preferred diatom species. Biomasses of 200 g/m^2 live-weight have been recorded. Length up to 200 mm.

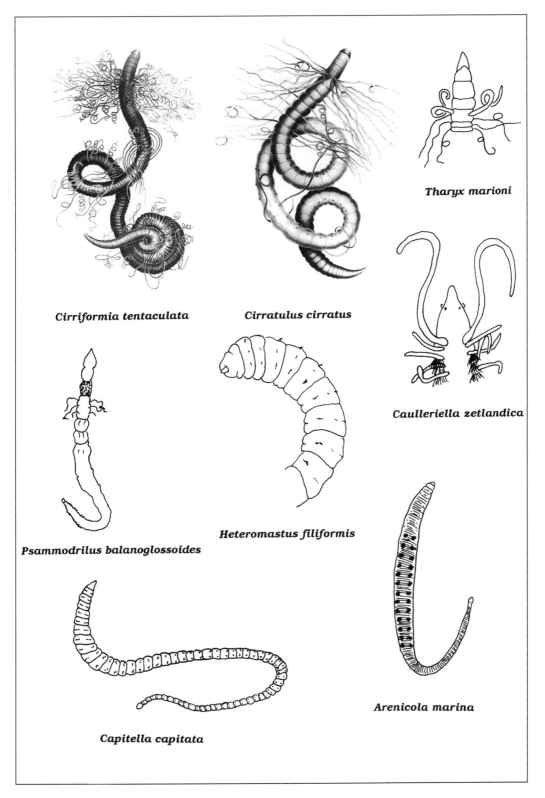

Fig. 43. Brackish-water polychaetes (VI).

Opheliida
Opheliidae

Armandia cirrhosa (Fig. 44)

A small, somewhat nematode-like, essentially Mediterranean species that was known in northwestern Europe only from gravelly sediment in a brackish pond in Hampshire, England, but which now appears to be extinct there. Almost nothing is known of its biology.

Ophelia bicornis (Fig. 44)

A southern species centred on the Atlantic coasts of France and Iberia that just extends into the southwest of the region. It is known only from clean, well-drained, high-salinity sands; those that it inhabits in the Exe Estuary, Devon, in densities of up to 360 individuals/m², support virtually no other macrofauna. There is a single record from non-tidal lagoon at a salinity of 24‰ (in Sussex). Length up to 50 mm.

Ophelia rathkei (Fig. 44)

O. rathkei can inhabit muddier sands than *O. bicornis* above, and can extend into estuaries to regions experiencing salinities down to 18‰. It occurs, often intertidally, around the British Isles and eastwards to Denmark. Length <10 mm.

Terebellida
Sabellariidae

Sabellaria alveolata (Fig. 44) and *S.spinulosa* (Fig. 44)

S. alveolata and the more sublittoral *S. spinulosa* construct thick-walled tubes of sand and (especially *S. alveolata*) can form large reefs. Neither species penetrates brackish waters to any great extent, but they may occur near the mouths of estuaries in waters of more than 22‰.

Ampharetidae

Alkmaria romijni (Fig. 44)

This small (up to 5 mm long) species is so far known only from lagoons and other non-tidal brackish habitats around the southern shores of the North Sea (England, Netherlands, Germany and Denmark) and part way down the English Channel (as far as Hampshire). It inhabits a mud tube to which the faecal pellets adhere. The young are benthic and are retained at first within the adult tube. The 8 tentacles are thread-like and slimy, and the 6 gills are banded by rings of greenish-grey pigment. A salinity of 4‰ appears to be the lower limit, and few records are from >25‰.

Melinna palmata (Fig. 44)

In a mud tube in *Zostera* meadows and muddy sediments generally; present throughout the region in the relatively marine sections of estuaries and in some lagoons. Length <50 mm.

Ampharete grubei (Fig. 44)

Inhabits a membraneous tube in muddy or sandy mud sediments, including in *Zostera* meadows, in the saltier lagoons and stretches of estuaries, although it occurs in waters of less than 18‰in some localities. Also present in the Baltic and from the Arctic to the Mediterranean. Frequently recorded as *Ampharete acutifrons*. Length <80 mm.

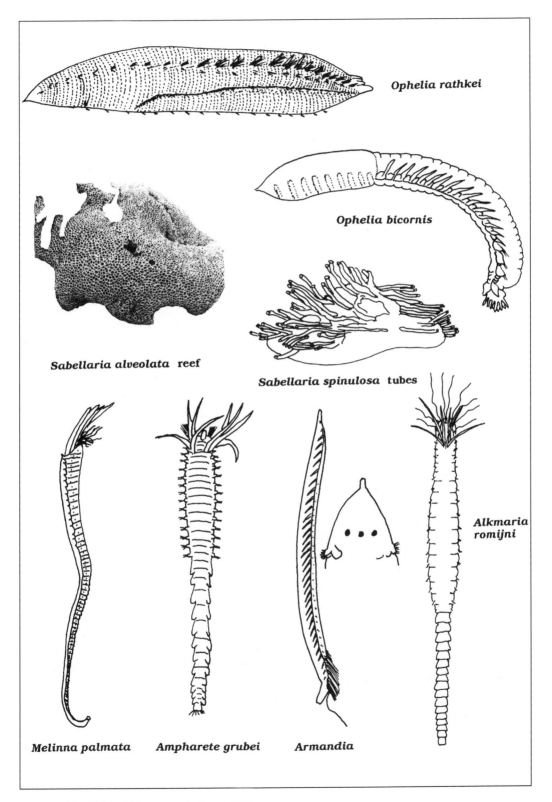

Fig. 44. Brackish-water polychaetes (VII).

Terebellidae
Amphitrite figulus (Fig. 45)
A widely distributed species that burrows into muds or muddy sands in sheltered waters, including in estuaries in salinities of down to some 22‰. Its long tentacles roam over the sediment surface and detrital food particles are conducted along the tentacles to the mouth. Length up to 250 mm.

Lanice conchilega (sand mason) (Fig. 45)
A widespread species with a distinctive sand and gravel tube that ends in a tassel projecting above the sediment surface. The animal can deposit-feed like *Amphitrite* above or the tentacles can be rested on the arms of the tassel and can then suspension-feed. The massed tubes may form low reefs. It does not occur in salinities of lower than 22‰.

Trichobranchidae
Terebellides stroemi (Fig. 45)
A euryhaline circum-Arctic species that occurs in the Baltic and south to the Mediterranean. It occupies a membraneous and sediment-covered tube in muddy, sandy or *Zostera* sediments, and – in the Baltic at least – can occur in salinities of down to some 3‰; elsewhere, however, its estuarine penetration appears to be considerably less. Length <90 mm.

Sabellida
Sabellidae
Fabricia stellata (Fig. 45)
Formerly known as *F. sabella*, this abundant and widespread, but easily overlooked, little worm occurs in mucus and detritus tubes in both tidal and non-tidal brackish sands and muds, and amongst *Chaetomorpha* and charophytes, down to salinities of some 4‰. Occasionally its massed tubes can form micro-reefs.

Fabriciola baltica (Fig. 45)
Superficially similar to *F. stellata*, this minute worm inhabits the surface layers of brackish-water muds in the Baltic in salinities of down to 8‰; it was recently discovered in Loch Etive, Scotland.

Manayunkia aestuarina (Fig. 45)
Superficially similar to the two species above, this minute species occurs over the whole brackish-water salinity range, down to <1‰, amongst *Ruppia* and other submerged vegetation, and in estuarine and lagoonal muds, including those in salt marshes. Their 1 cm long tubes (which it takes them 3 hours to construct) can be found throughout the region.

Sabella pavonina (Fig. 45)
A northeast Atlantic species, extending down into the Mediterranean, that occurs in the relatively salty areas of estuaries in most types of sediment, including those in *Zostera* meadows. It constructs a mud tube, often attached to a stone at depth, in waters of not less than 20‰. It has often been recorded as *S. penicillus*. Length up to 250 mm.

Myxicola infundibulum (Fig. 45)
A southern species, extending from the south of the region to the Mediterranean, that constructs a thick, transparent and gelatinous tube in estuarine muds down to salinities approaching 18‰. Length up to 200 mm.

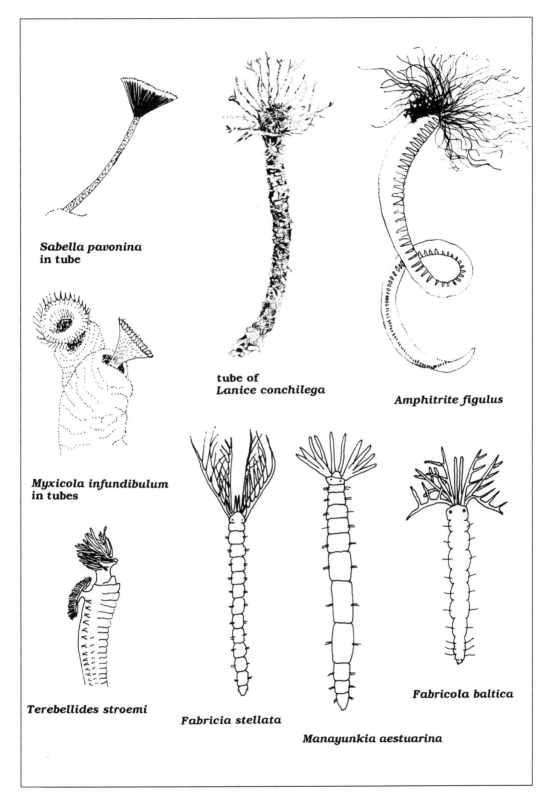

Sabella pavonina
in tube

tube of
Lanice conchilega

Amphitrite figulus

Myxicola infundibulum
in tubes

Terebellides stroemi

Fabricia stellata

Manayunkia aestuarina

Fabricola baltica

Fig. 45. Brackish-water polychaetes (VIII).

Serpulidae

Ficopotamus enigmatica (Fig. 46)

A southern species, often termed *Mercierella*, that extends northwards to the southern North Sea. It will construct its *c.* 1 mm diameter, up to 30 mm long tubes on most substrata, natural and artificial, including the stems of *Phragmites*, and is particularly characteristic of harbours, docks, artificial lagoons and power-station outfalls with salinities of from 35‰ down to 10‰ (and, exceptionally, in fresh water); it has also been recorded from several natural lagoons. The tubes, which possess circular, shelf-like platforms near their mouths, are often aggregated into large colonies.

Spirorbidae (Fig. 46)

A few species of spirorbids may be found in brackish waters, particularly on hard substrata in the mouths of estuaries. They are, however, impossible to identify whilst the animal is alive.

Clitellates: oligochaetes (tubifex and pot worms) and leeches

Small and often transparent versions of earthworms, which like their larger relatives live in the sediments and (except in asexual species) develop a clitellum over up to 6 of their otherwise uniform segments in the breeding season. Segmented worms like the polychaetes but never with appendages (tentacles or parapodia). See Brinkhurst (1982). One leech, *Helobdella stagnalis*, sometimes occurs in dilute brackish waters: it can be distinguished from other leeches by the presence of a small, chitinous scale about one sixth of the distance from the anterior to the posterior sucker, and by the single pair of unlobed eyes.

1 A Each bundle of chaetae containing only 2 chaetae .... lumbriculids
  B Each bundle of chaetae containing more than 2 chaetae... 2

2 A Whitish, particularly earthworm-like, species with chaetae in lateral and ventral bundles; without hair-like chaetae .... enchytraeids
  B Delicate, transparent, thin-cuticled species with chaetae in dorsal and ventral bundles; often with long hair-like chaetae 3

3 A Small (<20 mm long); often with eyespots; reproduce by budding to form chains of individuals.. naidids
  B Larger (>20 mm long); without eyespots; never forming chains of individuals; often reddish .... tubificids

CLITELLATA: OLIGOCHAETA
Lumbriculida
Lumbriculidae (Fig. 46)

A family (and order) containing the single species *Lumbriculus variegatus:* a bright red-brown, up to 40 mm long, earthworm-like species that swims with spiral undulations and occurs in most types of

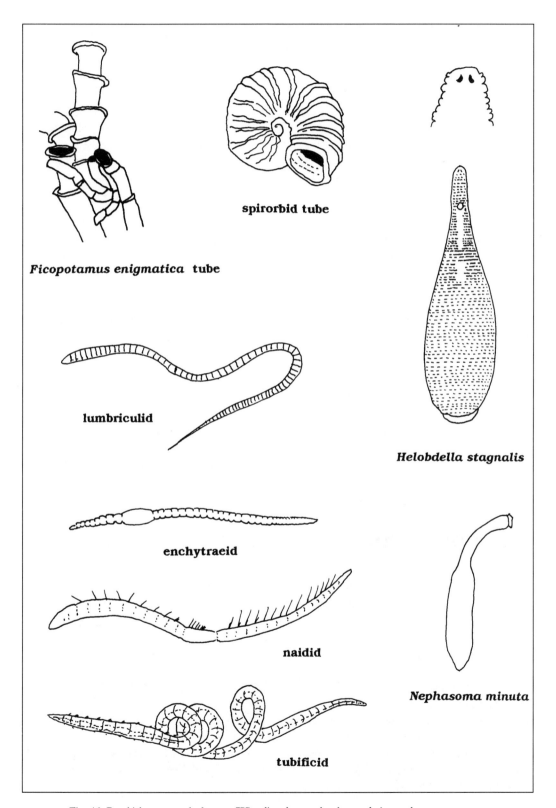

Fig. 46. Brackish-water polychaetes (IX), oligochaetes, leeches and sipunculans.

sediment and amongst submerged vegetation. It is essentially a fresh water species that can occur in some low-salinity ponds (<8‰), especially along the southern shores of the North Sea.

Haplotaxida
Naididae (Fig. 46)

The small, delicate naidids are mostly fresh water in habitat, but species of *Chaetogaster, Nais, Paranais* and *Stylaria,* together with *Amphichaeta sannio* and *Uncinais uncinata,* occur in relatively dilute lagoons and equivalent pools in Britain and along the southern North Sea coast, as well as in the Baltic. They occur in algal turfs and fine sediment. Asexual multiplication by budding off chains of individuals is the norm. *Chaetogaster* is unusual amongst oligochaetes in being at least partly predatory.

Tubificidae (Fig. 46)

Tubificids are abundant in the sediments of most estuaries and non-tidal lagoons. Characteristic genera include *Tubifex, Limnodrilus, Potamothrix, Tubificoides* (=*Peloscolex*), *Clitellio, Monopylephorus* and *Aktedrilus.* Densities can attain almost 500 000 individuals/m² and dry-weight biomasses 80 g/m². Between the various species, the whole salinity range from fresh water to the sea is covered.

Enchytraeidae (Fig. 46)

The relatively earthworm-like enchytraeids are mainly soil-dwelling, but species of *Grania, Marionina* and especially *Lumbricillus* (not to be confused with *Lumbriculus* above) occur in intertidal estuarine sediments, and *Lumbricillus* has been recorded from non-tidal brackish pools.

CLITELLATA: HIRUDINEA
Rhynchobdellida
Glossiphoniidae

Helobdella stagnalis (Fig. 46)

This leech has been recorded from a few low-salinity lagoons and equivalent habitats, in salinities of less than 8‰. Up to 10 mm long when at rest.

Sipunculans

The only sipunculan likely to be encountered in brackish water, *Nephasoma minuta*, is a small (<15 mm long) cylindrical, sausage-shaped, unsegmented worm with a long anterior end (the 'introvert'), tipped by fleshy lobes that can be extruded from the body or withdrawn into it. It occurs in lagoonal shingle at the water line. Gibbs (1977) describes British species.

Golfingiidae

Nephasoma minuta (Fig. 46)

This small hermaphrodite sipunculan, no more than 15 mm long, occurs – in brackish water – in the shingle enclosing lagoons from Dorset to East Anglia, England, specifically in areas where the sea water percolates through and the salinity is above 20‰. It is otherwise known from soft sediments, crevices and amongst *Sabellaria* tubes throughout the North Sea and Channel region, mostly under the generic name *Golfingia.*

Molluscs

Chitons (coat-of-mail shells)

Flat, slug-like molluscs, up to 30 mm long, bearing on the upper surface a series of 8 transverse shelly plates surrounded by a tough 'girdle' that completely hides the head and foot. Alongside the foot are many pairs of gills. Two species may occur on shingle or shell debris: *Leptochiton asellus* with a very narrow girdle and gills only in the posterior half of the body; and *Lepidochitona cinereus* with an extensive girdle and gills along almost the whole length of the body. Jones & Baxter (1987) describe British species.

POLYPLACOPHORA
Lepidopleurida
Lcpidopleuridae
Leptochiton asellus (Fig. 48)
A common chiton found attached to stones and shell debris from the Arctic to Spain, including in Scottish lagoons and The Fleet, Dorset. <20 mm long.

Ischnochitonida
Ischnochitonidae
Lepidochitona cinereus (Fig. 48)
A common chiton, usually found on stones, from the Arctic to the Mediterranean, including in estuaries and shingle-floored lagoons. <30 mm long.

Shelled gastropods (snails, winkles, whelks, etc.)

Molluscs of up to 10 cm shell height, although more usually <1 cm, with an external, single, conically spiral, hard shell (a few limpet-shaped species may also occur) and often with an operculum that can close the aperture of the shell on withdrawl of the snail. In all brackish-water habitats, except swimming in the water. Graham (1988) and Thompson (1988) describe British species. See Fig. 47.

1 A Body considerably larger than shell and cannot withdraw into it when disturbed; pale amber shell visible only at the posterior end of the body, anteriorly the grey – dark-brown mottled body extends over the front of the shell in the form of a pair of lateral flaps or 'wings' .... *Akera bullata*
  B Body, even if partially covering the shell, can withdraw into it when disturbed 2

2 A Aperture of the shell can be plugged by an operculum after withdrawl of the animal, the operculum being carried dorsally on the posterior part of the foot when the animal is crawling, or animal limpet-shaped 9

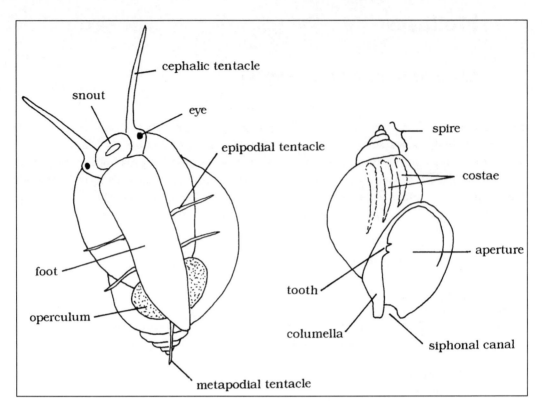

Fig. 47. Anatomical features used in the identification of shelled gastropod molluscs.

B Operculum absent and animal not limpet-shaped 3

3 A Rear end of the foot forked .... *Diaphana minuta*
  B Rear end of foot not forked 4

4 A Shell fragile, without an obvious spire, partially covered by parts of the body in life, with
    an elongate aperture extending the whole or almost the whole height of the shell 5
  B Shell sturdy, with a distinct spire and with the aperture extending a distance <shell height;
    shell not partly covered by folds of the body in life 7

5 A Head without a pair of posterolateral tentacles; dark-coloured body may exceed 50 mm
    length; shell almost circular in outline .... *Haminea navicula*
  B Head with a pair of posterolateral tentacles; whitish body <15 mm long; shell elongate,
    almost parallel sided 6

6 A Cephalic tentacles rounded; aperture of shell shorter than shell height .... *Retusa obtusa*
  B Cephalic tentacles pointed; aperture of shell longer than shell height .... *Retusa truncatula*

7 A Aperture of shell without teeth on the columella .... *Lymnaea* spp
    *[This group of fresh water snails occasionally penetrates into the most dilute of brackish waters. The most likely species
    so to do is* L. peregra *– this has a short spire and a large aperture.]*
  B Aperture of shell with at least 2 teeth on the columella 8

8 A Aperture of shell with 2 teeth on the columella but without any on the outer lip
.... *Leucophytia bidentata*

B Aperture of shell with 3 or more teeth on the columella and with the same number on the
outer lip .... *Ovatella myosotis*
[This species is often termed Phytia myosotis*]*

9 A Shell neither spirally coiled nor limpet-shaped, but forming a small, short, slightly curved
tube open at one end and closed at the other (restricted to lagoonal shingle at the water
level) .... *Caecum armoricum*

B Shell either spirally coiled or limpet-shaped 10

10 A Shell limpet-shaped, not spirally coiled although the apex may be curved......................... 11

B Shell spirally coiled 14

11 A Shell with a radial slit extending from the edge almost to the apex; often pinkish, up to 6
mm long and 4 mm high... *Emarginula conica*

B Shell without any slit up its side 12

12 A Shell smooth, with chocolate brown reticulate pattern radiating from apex; with a single gill
near the head .... *Collisella tessulata*

B Shell without chocolate markings; with many gills along the sides of the foot 13

13 A Shell smooth with blue rays; without gills anteriorly .... *Helcion pellucidum*

B Shell with radiating ridges; with gills all around the mantle .... *Patella vulgata*

14 A Aperture of shell drawn out anteriorly into a canal or tube through which the siphon can
be extended 15

B Aperture of shell entire anteriorly, without any siphonal notch, canal or tube 17

15 A Posterior end of foot with a V-shaped notch and 2 tentacles; eye only a little above base of
cephalic tentacle; animal brown-black with white speckling; shell reticulate .... *Hinia reticulata*

B Posterior end of foot not notched nor with tentacles; eye one third of cephalic tentacle
length up from base; animal light-coloured although maybe with dark markings 16

16 A Siphon long; shell usually yellowish brown, up to 10 cm high, marked with spiral ridges
and reversed-C shaped costae (the edible whelk) .... *Buccinum undatum*

B Siphon short; shell usually whitish or grey, sometimes with brown bands, with spiral ridges
but without costae; aperture often thickened (the dogwhelk).......................... *Nucella lapillus*

17 A Shell narrow and elongate, >3× as tall as its greatest breadth, the sides subtending an angle
<30° 18

B Shell not so elongate, 3× or less as tall as its greatest breadth, the sides subtending an angle
>30° 20

18 A Aperture with flared lips; body cream to light brown; cephalic tentacles white, elongate
.... *Rissostomia membranacea/labiosa*

B Aperture without flared lips; without the combination of pale body and elongate cephalic
tentacles.. 19

19 A Body pale, head with 2 triangular tentacles between the bases of which are the eyes; shell white, with broad costae .... *Turbonilla lactea*

   B Body dark (brown streaked with black & white), head with long slender cephalic tentacles; shell light brown, reticulated.. *Bittium reticulatum*

20 A Shell a small (up to 1.5 mm across) brown, flat or almost flat disc (like an ammonite or ramshorn), twice as broad as high; aperture circular.. 21

   B Shell conical, not a flat disc... 22

21 A Shell coiled in a single plane, forming a biconcave disc.............................. *Omalogyra atomus*

   B Shell coiled helically, upper surface slightly convex, lower surface concave..... *Skeneopsis planorbis*

22 A Shell with a very low or without a distinct spire, the last whorl forming >90% of total shell height.. 23

   B Shell with a spire comprising at least 15% shell height 27

23 A Aperture of shell D-shaped.. 24

   B Aperture of shell oval or circular 25

24 A Shell solid, dark with light-coloured streaks, without an umbilicus; eyes on stalks alongside the cephalic tentacles.. *Theodoxus fluviatilis*

   B Shell fragile, greenish, with a wide umbilicus; eyes at base of the cephalic tentacles .... *Lacuna pallidula*

25 A Shell thin, taller than broad, aperture forming <70% of shell height; shell dark brown to black often with a reticulate pattern on the last whorl; associated with tassel weed (*Ruppia*) or filamentous green algae... *Littorina saxatilis* var. *lagunae*

   B Shell solid, as broad or broader than tall, aperture forming at least 75% of shell height; shell colour various (including almost black); associated with fucoid seaweeds 26

26 A Low spire present (forming 5–10% of shell height); aperture wide, comprising 75–80% of shell height; associated with the seaweeds *Fucus vesiculosus* and *Ascophyllum nodosm* .... *Littorina obtusata*

   B Spire forming 0–5% of shell height; aperture narrow with a constricted throat, forming 85–90% of shell height; associated with the seaweed *Fucus serratus* .... *Littorina mariae*

27 A Shell as broad (to within 10%) or broader than tall 28

   B Shell taller than broad.. 32

28 A Shell small, glossy, semitransparent, as broad as high; animal with bifid snout, without epipodial tentacles 29

   B Shell not semitransparent, as broad or broader than high; animal with paired epipodial tentacles arising at intervals along its foot, snout not bifid... 30

29 A Eyes located close to cephalic tentacle tips; spire forming some 25% of shell height; occurs at or above the water line .... *Paludinella littorina*

   B Eyes behind the cephalic tentacle bases; spire forming <20% of shell height; occurs below water level .... *Rissoella globularis*

30 A With 6 pairs of epipodial tentacles; shell small (<6 mm across), glossy, horn-coloured, with green and purple highlights... *Margarites helicinus*

B With 3 pairs of epipodial tentacles; shell up to 20 mm or more across, light-coloured and with radiating bands of reddish or brownish purple 31

31 A Colour pattern of many fine bands each <0.5 mm wide; umbilicus small and egg-shaped; shell relatively tall, about as high as broad .... *Gibbula cineraria*

B Colour pattern of few broad bands each >1 mm wide; umbilicus large and circular; shell relatively low, 1.3–1.4 × as broad as tall .... *Gibbula umbilicalis*

32 A Shell cylindrical, as wide at the apex as near the aperture in adults, buff-coloured, 5 mm high; animal with cylindrical snout ending in a rounded mouth disc; occurring at the water level .... *Truncatella subcylindrica*

B Shell narrowing towards the apex.. 33

33 A Shell with costae aligned more or less parallel to its long axis 34

B Shell without costae 40

34 A Last shell whorl with a reticulate series of pits in its surface, up to 5 mm high, translucent; costae visible on only one whorl when viewed from aperture........................ *Rissoa rufilabrum*

B Last shell whorl without such pits.. 35

35 A Shell breadth at least half shell height 36

B Shell breadth <half shell height.. 38

36 A With dark comma-shaped mark on the last whorl near the aperture; shell up to 5 mm high; body pale yellow with dark streaks, cephalic tentacles pale with central white line, with yellow patch behind each eye .... *Rissoa parva*

B Last shell whorl without dark comma-shaped streak; shell up to 3 mm tall 37

37 A Apex of shell bronze; with flat metapodial tentacle .... *Pusillina sarsi*

B Apex of shell purple; with filiform metapodial tentacle; costae numerous but fine
.... *Pusillina inconspicua*

38 A Shell up to 10 mm high, without spiral ridges and grooves; costae extending right across whorls .... *Rissostomia membranacea/labiosa*

B Shell up to 4 mm high, with numerous spiral ridges and grooves, and with costae only on the apical halves of whorls... 39

39 A Costae present on all whorls.. *Onoba semicostata*

B Costae only on apical whorls .... *Onoba aculeus*

40 A Shell with spiral ridges and grooves and/or with spiral bands of brown pigment................ 41

B Shell without any form of surface sculpture, apart from growth lines; nor more than 1 spiral band of brown pigment 47

41 A Shell with a single spiral ridge or row of bristles .... *Potamopyrgus antipodarum*
[This species was for many years known as Hydrobia, *and later as* Potamopyrgus, jenkinsi *]*

B Spiral ridges and/or pigment bands more than one 42

42 A Shell breadth >half shell height.. 43
B Shell breadth half or less than half shell height 44

43 A Outer lip of aperture arises tangentially from the shell; cephalic tentacles with transverse dark stripes; columella white (the edible winkle) .... *Littorina littorea*
B Outer lip of aperture arises at right angles to the last whorl; cephalic tentacles with lateral longitudinal stripes; columella dark .... *Littorina saxatilis*

44 A Shell with spiral bands of dark pigment 45
B Shell without spiral bands of dark pigment 46

45 A With 3 brown bands on last whorl (with various degrees of fusion); with single short, triangular metapodial tentacle; columella without groove leading to umbilicus; up to 4 mm tall .... *Cingula trifasciatus*
B With 4 orange-brown bands on last whorl (with various degrees of fusion); 2 metapodial tentacles project from under operculum; columella with groove leading to umbilicus; up to 10 mm tall .... *Lacuna vincta*

46 A Posterior end of foot with median cleft; without a metapodial tentacle; shell semitransparent, white but appearing yellow-orange because of the periostracum .... *Ceratia proxima*
B Posterior end of foot not cleft; with flat metapodial tentacle projecting from under operculum; shell not semitransparent .... *Onoba aculeus*

47 A Cephalic tentacles triangular, joined together at their bases, each with marked lateral groove; with a forwardly-projecting area of the foot (the mentum) and elongate proboscis; eyes between the tentacles; larval shell tucked upside down into the postlarval whorls 48
B Cephalic tentacles, eyes and larval shell not as above; without mentum or proboscis............ 49

48 A Whorls tumid; columella lip of aperture without a noticeable tooth; parasitic/predatory on *Mytilus*, etc.. *Brachystomia rissoides*
B Whorls flat-sided; columella lip of aperture with prominent tooth; parasitic/predatory on *Crassostrea*, etc. .... *Brachystomia eulimoides*

49 A The only tentacles on the head forming short, rounded lobes, each with a large eye at its tip; shell squatly conical, tan-coloured.. *Assiminea grayana*
B Cephalic tentacles elongate, filiform; eyes near or at bases of these tentacles...................... 50

50 A Cephalic tentacles bifid, each tentacle forming a V .... *Rissoella opalina*
B Cephalic tentacles simple 51

51 A Cephalic tentacles banded transversely in black; shell large, solid, straight sided (edible winkle).. *Littorina littorea*
B Cephalic tentacles without transverse banding 52

52 A Shell elongate, with up to 9 whorls, almost 3× as tall as broad; aperture with flared lips; cephalic and metapodial tentacles white .... *Rissostomia membranacea/labiosa*

B Aperture without flared lips; shell not as elongate as above 53

53 A Cephalic tentacles each with a transverse black mark close to the tip............................... 54
B Cephalic tentacles without such a black mark near the tip 55

54 A Black mark in the form of a marked rectangular patch; shell solid, brownish, straight-sided and squatly conical when intertidal, more fragile, darker and tumidly whorled when lagoonal .... *Hydrobia ulvae*
B Black mark in the form of a less distinct, inverted V or a cone; shell almost black, shiny, somewhat parallel sided .... *Hydrobia neglecta*

55 A Shell forming a squat cone, its sides subtending an angle of >70°................................... 56
B Shell forming a taller cone, its sides subtending an angle of <70°.................................. 57

56 A Shell up to 12 mm tall,with a prominent and acute spire comprising 15–30% of shell height, and with projecting columellar lip; occurring on the substratum
.... *Littorina saxatilis* var. *tenebrosa*
B Shell <6 mm tall, with depressed and rounded spire comprising 7–17% of shell height; without a projecting columellar lip; occurring on submerged macrophytes and algae
.... *Littorina saxatilis* var. *lagunae*

57 A Shell milky-white, glossy, minute (<1.5 mm height), with blunt apex; dark oval mark on body visible through the semitransparent shell; operculum with T-shaped brown mark
.... *Rissoella diaphana*
B Shell and operculum not as above... 58

58 A Body light coloured (white to yellow, with or without darker markings), with metapodial tentacle.. 59
B Body colour dark (grey to black, with or without lighter patches), without a metapodial tentacle 60

59 A Metapodial tentacle flat; shell up to 3 mm high, tumidly whorled....................... *Pusillina sarsi*
B Metapodial tentacle filiform; shell up to 5 mm high, straight sided.... *Rissoa parva* (var. *interrupta*)

60 A Shell cone relatively squat, the sides subtending an angle of >45°, the last whorl large, forming some 70% of the shell height 61
B Shell taller, the sides subtending an angle of <45°, the last whorl forming some 60% of shell height 62

61 A Shell with umbilicus; cephalic tentacles pale and with median longitudinal brown band; orange-yellow spot behind each eye .... *Pseudamnicola confusa*
B Shell without umbilicus; cephalic tentacles blue-grey at the sides (and appearing crinkled) and with a narrow median longitudinal pale line .... *Potamopyrgus antipodarum*
[This species was for many years known as Hydrobia, *and later as* Potamopyrgus, jenkinsi]

62 A Penis* with a long terminal flagellum, without lateral bulbous protruberances; cephalic tentacles uniformly coloured (usually translucent) although sometimes with a thin median dark line.. *Hydrobia ventrosa*

B Penis* without a terminal flagellum, with 5 or so bulbous protruberances issuing near the base and another near the tip; cephalic tentacles dark laterally but with a median longitudinal light band and a light tip .... *Heleobia stagnorum*
[*Recorded as* Hydrobia, *or* Semisalsa, stagnorum]

*Only the males of these two species are readily distinguishable: the penis is located behind the cephalic tentacles and can be viewed by inducing a male to extend its head and 'neck' out of its shell, e.g. by (gently!) holding the shell slightly above the bottom of a petri-dish whilst observing through a microscope.

Shell-less gastropods (sea-slugs)

Aquatic slugs (in effect shell-less snails, although some have a shell enclosed by body lobes) often with the upper surface of the body drawn out into a series of soft cerata (Fig. 56), or with a circlet of gills towards the rear end of the body, or with two, projecting, anterior, sensory organs looking something like ears. Crawl on a stout, muscular foot on the sediment/shingle or more commonly over surfaces, including the vegetation, supporting hydroids or bryozoans on which they feed. Most brackish-water species are small (<10 mm long), although a few may exceed 10 cm. Thompson (1988) describes British species.

1 A Dorsal surface of animal bearing a pair of anterior tentacles and a circlet of gills or a series of tentaculate or leaf-like lobes or any other series of large projections 2
  B Dorsal surface of animal without any projecting lobes or circlets of gills, except for tentacles on the head or for the tips of gills that may protrude from the mantle cavity on the right-hand side posteriorly 15

2 A Dorsal projections include a series of finger-, lobe- or leaf-like structures (cerata) along the body; without a partial or complete circlet of gills centrally or posteriorly 3
  B Dorsal projections in the form of a pair of stout, anterior, tentacle-like rhinophores and a partial or complete circlet of gills (which may be retractile) in the middle or posterior region of the body; without cerata 9

3 A Without anterior tentacles; occurring on mats of cyanobacteria in salt-marshes .... *Alderia modesta*
  B With at least 1 pair of anterior tentacles 4

4 A With 2 pairs of anterior tentacles 5
  B With 1 pair of anterior tentacles 6

5 A Cerata with 2–3 brown to olive green rings; tips of tentacles white .... *Eubranchus exiguus*
  B Cerata blotched, but not ringed, by brown to olive green pigment; tips of tentacles paler than rest but not white .... *Eubranchus farrani*
[*It is possible that various* Aeolidiella *spp. will turn up in high-salinity brackish waters, feeding on their sea-anemone prey. They can be distinguished from* Eubranchus *by their possession of more than 100 cerata. Further, it is conceivable that* Facelina *spp. might also occur: they can be distinguished from both* Eubranchus *and* Aeolidiella *by the occurrence of three pairs of anterior tentacles.*]

6 A Cerata with whorls of large, rounded tubercles, each tubercle with a small crimson spot
.... *Doto coronata*

B Cerata smooth, without tubercles 7

7 A Cerata swollen, rounded; with a pair of prominent eyes behind the tentacles .... *Stiliger bellulus*

B Cerata elongate, finger-like; eyes not obvious (at base of tentacles) 8

8 A Anterior tentacles inrolled; cerata greenish or brownish; without a dome-shaped oral hood
.... *Hermaea dendritica*

B Anterior tentacles not inrolled; cerata pale yellow – orange – pink; with dome-shaped oral
hood .... *Tenellia adspersa*
[This species has often been recorded under the name Embletonia pallida*]*

9 A With only 3 gills forming an anterior half-circle; body with numerous large club-shaped
tubercles each ending in a red or black spot, and with a number of brilliant blue pigment
spots on the body surface .... *Aegires punctilucens*

B Gills forming a complete circle; body without large club-shaped tubercles 10

10 A Body elongate, with a distinct dorsal rim running around the body from in front of the
rhinophores to behind the gill circlet, from beneath the rim protrude a posterior tail and a
pair of anterior flattened oral tentacles; body a translucent white, the rhinophores being
yellowish .... *Goniodoris nodosa*

B Body an oval disc without a dorsal rim or visible oral tentacles (the tip of the tail may protrude from beneath the disc when crawling) 11

11 A Gill circle, if touched, retracts *en masse* into a cavity within the body 12

B Gills, if touched, individually contract but do not disappear *en masse* within the body 13

12 A Body translucent white, flattened; up to 30 mm long and with up to 7 gills .... *Cadlina laevis*

B Body with blotchy markings in a variety of colours (red, yellow, brown, green, etc.); up to
120 mm long; with 8 or more pale gills .... *Archidoris pseudoargus*

13 A Central region of dorsal surface with elongate soft tentacles; body up to 50 mm long; with
up to 9 large tripinnate gills .... *Acanthodoris pilosa*

B Central region of dorsal surface with large rounded tubercles; body up to some 20 mm
long; with up to 12 small, simply pinnate gills 14

14 A Body yellow-orange (sometimes paler and may be white in northern regions); rhinophores
and gills darker than rest of body.. *Adalaria proxima*

B Body white (occasionally pale yellow, especially in northern regions); rhinophores and gills
colourless .... *Onchidoris muricata*

15 A Head with a pair of finger-like tentacles... 16

B Head without finger-like tentacles 19

16 A Body oval or circular, without a pair of conspicuous eyes behind the tentacles.................. 17

B Body elongate; with a pair of conspicuous eyes behind the tentacles 18

17 A Body oval, up to 60 mm in length, yellow to orange in colour without black patches; with a single gill protruding from the mantle cavity on the right-hand side [this species does in fact have a shell but it is not externally visible]....................................... *Berthella plumula*

   B Body circular, some 5 mm long, with black patches or entirely dark coloured; with 5 gills protruding from the mantle cavity on the right-hand side............................. *Doridella batava*

18 A Body <6 mm long, brownish or black; without a pair of lateral 'wings'; tentacles simple, not inrolled.. *Limapontia senestra*

   B Body up to 45 mm long, green to red, depending on diet; with a pair of large, mobile, wing-like lobes of the body running down from behind the head to the posterior tip; tentacles inrolled .... *Elysia viridis*

19 A With 3 simple gills posteriorly that may be seen projecting out of the mantle cavity .... *Runcina coronata*

   B Without gills 20

20 A With ear-like lobes above the eyes; not associated with salt-marsh mud; excretory pore and anus close together.. *Limapontia capitata*

   B Without any lobes above the eyes; associated with salt-marsh mud; excretory pore and anus not close together.. *Limapontia depressa*

GASTROPODA: PROSOBRANCHIA
Pleurotomariida
Fissurellidae
Emarginula conica (Fig. 48)
This tiny (<6 mm diameter) limpet occurs on stones and shingle in The Fleet, Dorset. It is otherwise known from Scotland and Eire to the Mediterranean.

Docoglossida
Acmaeidae
Collisella tessulata (Fig. 48)
An essentially northern species that occurs in some of the rocky or stony brackish-water lochs of Scotland and Denmark, but not elsewhere. <20 mm diameter.

Patellidae
Patella vulgata (common limpet) (Fig. 48)
The common limpet may be found in estuaries and estuarine lagoons on hard substrata, whether natural or artificial, down to salinities of some 20‰. <50 mm diameter.

Helcion pellucidum (blue-rayed limpet) (Fig. 48)
Occurs in some of the rocky brackish-water lochs of Scotland, but not elsewhere, presumably because of the lack of appropriate substrata. <20 mm diameter.

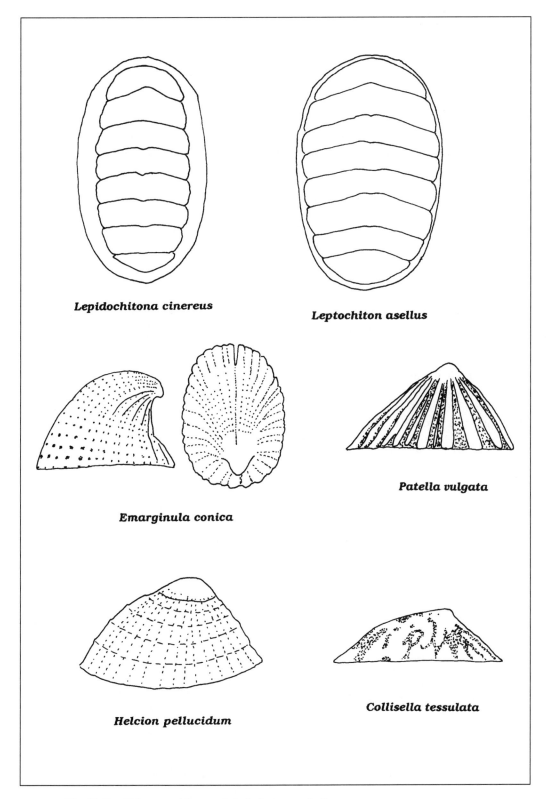

Fig. 48. Brackish-water chitons and shelled gastropods (I).

Anisobranchida
Trochidae
Margarites helicinus (Fig. 49)
Occurs in several of the rocky brackish-water lochs of Scotland, but not elsewhere. It is a circumboreal species.

Gibbula cineraria (grey topshell) (Fig. 49)
Occurs in several of the rocky brackish-water lochs of Scotland and in those few estuaries (for example Milford Haven) where rocks occur in salinities of more than 20‰, but not elsewhere, presumably because of the lack of appropriate hard and/or algal substrata. <17 mm tall.

Gibbula umbilicalis (purple topshell) (Fig. 49)
A generally western species, absent from the North Sea. It occurs in a few rocky brackish-water lochs of Scotland and rarely in other brackish waters in the south west. <17 mm tall.

Neritida
Neritidae
Theodoxus fluviatilis (Fig. 49)
Essentially a fresh water species that can withstand salinities of up to some 15–20‰ in and near the Baltic, and that occurs in lagoon-like habitats along the southern shore of the North Sea. Up to 8 mm tall and broad.

Neotaenioglossa
Lacunidae
Lacuna pallidula (Fig. 49)
Occurs in some of the rocky brackish-water lochs of Scotland, and rarely elsewhere. Mainly associated with *Fucus*, but sometimes on *Zostera* , in waters of >15‰. Up to 12 mm tall.

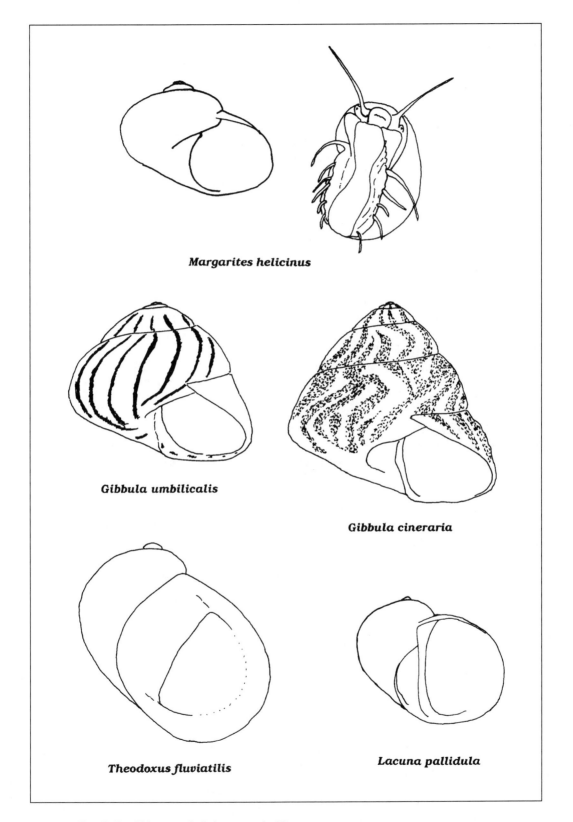

Margarites helicinus

Gibbula umbilicalis

Gibbula cineraria

Theodoxus fluviatilis

Lacuna pallidula

Fig. 49. Brackish-water shelled gastropods (II).

Lacuna vincta (Fig. 50)
Occurs in some of the rocky brackish-water lochs of Scotland, and more rarely on *Zostera* elsewhere. Able to withstand salinities down to some 15‰ in Denmark. Up to 10 mm tall.

Littorinidae (winkles)
Littorina saxatilis (rough winkle) (Fig. 50)
A common inhabitant of estuarine salt-marshes and gravelly areas of mudflats, and also found around the shingle margins of lagoons, especially those near the sea or formed relatively recently. 8‰ appears to be its lower limit. Up to 15 mm tall.

Littorina saxatilis var. *tenebrosa* (Fig. 50)
A high-spired form of *L. saxatilis* that occurs on the beds of old, well-established lagoons and in other tideless habitats, such as areas of the Baltic.

Littorina saxatilis var. *lagunae* (Fig. 50)
A small, globular and probably paedomorphic member of the *L. saxatilis* aggregate that occurs permanently submerged on vegetation such as *Ruppia* in lagoons, down to salinities of 8‰. It appears not to descend to the substratum except as a result of weed die-back.

Littorina littorea (edible winkle) (Fig. 50)
Associated with gravelly areas of estuarine mudflats in relatively high salinity regions (adults can withstand 10‰, but the eggs require double this to develop). Occurs from the Arctic southwards to Spain. Up to 30 mm tall.

Littorina obtusata (Fig. 50)
Occurs wherever its *Fucus vesiculosus* and/or *Ascophyllum nodosum* (and occasionally *Zostera*) food is found; that is most commonly in the relatively rocky north of the region. A specific brackish-water form found in the estuaries of East Anglia, with a higher spire and narrower aperture than usual, was described as *L. aestuarii* – few regard it as a separate species, and in any event it may now be extinct. Up to 16 mm tall and broad.

Littorina mariae (Fig. 50)
Occurs in many of the rocky brackish-water lochs of Scotland, but not elsewhere, presumably because of the lack of appropriate hard substrata with *Fucus serratus*. Up to 12 mm tall and broad.

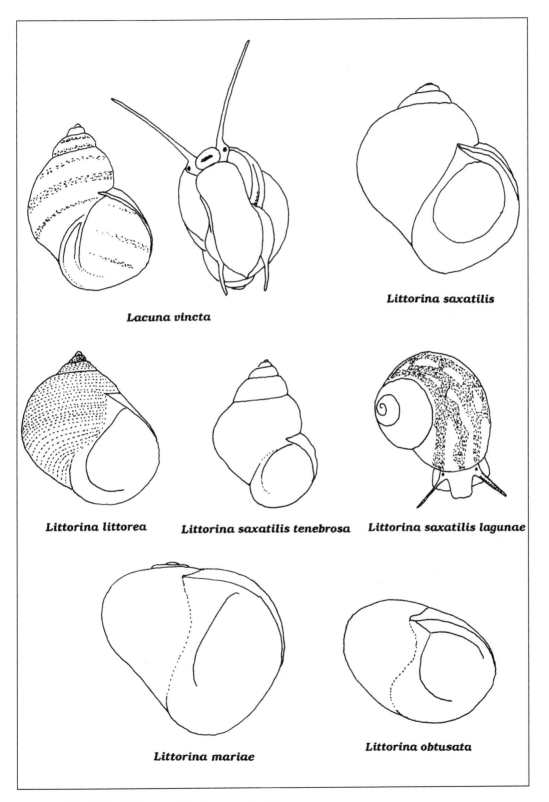

Fig. 50. Brackish-water shelled gastropods (III).

Hydrobiidae (mudsnails)

Hydrobia ulvae (Fig. 51)

Widespread on estuarine mudflats and salt-marshes down to salinities of some 4‰, occurring in densities of up to 300 000 individuals/m^2 and feeding mainly on diatoms. Also present in some lagoons (often together with other hydrobiid species), where it grows larger than intertidally (occasionally >8 mm tall), but much more characteristically intertidal in habitat (where rarely >5 mm tall). Unlike other hydrobiids it has a veliger larva. Occurs from the Arctic to West Africa and allegedly (though doubtfully) in the Mediterranean. The staple food of numerous invertebrates and vertebrates.

Hydrobia neglecta (Fig. 51)

A rare and sporadic *Hydrobia* that outside Denmark occurs in only a handful of localities from Eire in the west and Scotland in the north to Brittany in the south. With one exception restricted to lagoonal habitats, and usually in salinities >10‰. May be synonymous with *H. minoricensis* which may itself be synonymous with the common Mediterranean *H. acuta*. Up to 6 mm high.

Hydrobia ventrosa (Fig. 51)

A widespread species largely replacing *H. ulvae* in tideless lagoons and in the Baltic, and occurring over the whole brackish-water salinity range. Besides northwestern Europe it occurs around the shores of the Mediterranean and Black Seas, and possibly more widely (e.g. if, as some think, it is the same as the American *H. truncata* and the same as one of the hydrobiids in the Caspian). Virtually all continental records of *H. stagnorum* are really of this species (see *Heleobia stagnorum* below). Up to 6 mm high.

Potamopyrgus antipodarum (Fig. 51)

For many years known as *Hydrobia*, and later *Potamopyrgus, jenkinsi*, but now regarded as being identical to the New Zealand *P. antipodarum*, it was introduced to western Europe in the middle of the last century. In Europe, it is mainly fresh water in habitat, although it flourishes in lagoons in salinities of up to 20‰ and also occurs in some estuaries. Most populations are parthenogenetic; all are ovoviviparous. Up to 6 mm high.

Heleobia stagnorum (Fig. 51)

As a result of confusion with *Hydrobia ventrosa*, the distribution of this species is poorly known. With certainty, living populations within the region are known only from The Netherlands, although subfossil material is present along the Channel coast of England, and it occurs in the Mediterranean. Virtually all continental records of *Hydrobia stagnorum* therefore refer to *H. ventrosa*. Its generic position is unsettled, it having previously been assigned (besides *Hydrobia*) to *Semisalsa*.

Pseudamnicola confusa (Fig. 51)

A relatively rare species confined to the nearly fresh reaches of estuaries in the southern half of Eire, in southeastern England and in the Low Countries, where it occurs near the high water mark. It is common in southwestern Europe. Up to 4 mm high.

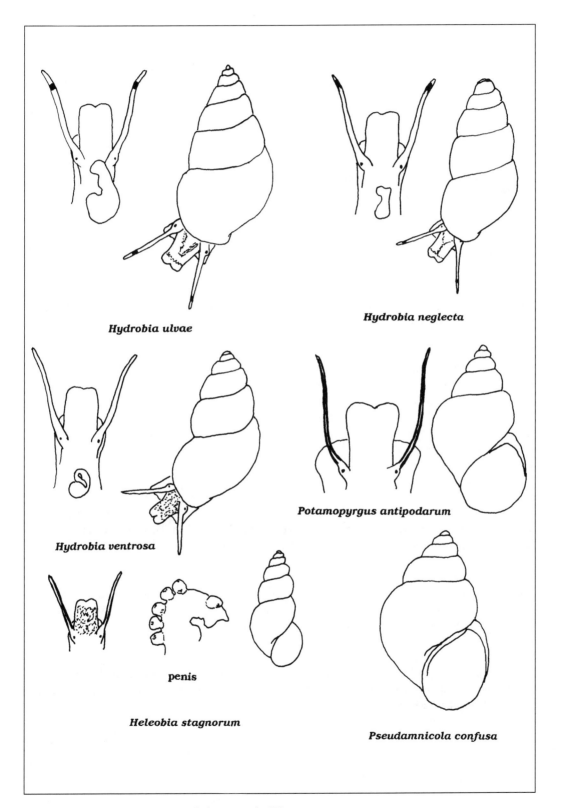

Hydrobia ulvae

Hydrobia neglecta

Hydrobia ventrosa

Potamopyrgus antipodarum

penis

Heleobia stagnorum

Pseudamnicola confusa

Fig. 51. Brackish-water shelled gastropods (IV).

Truncatellidae

Truncatella subcylindrica (Fig. 52)

An essentially southern species, now abundant in the region only beneath vegetation in the shingle enclosing The Fleet, Dorset, and more rarely there under stones, where it is found feeding on detritus at the water level. Has a peculiar looping gait.

Iravadiidae

Ceratia proxima (Fig. 52)

A small (<3 mm high) uncommon snail that occurs in low densities in the soft sediments of a few lagoons. Occurs from the Western Isles of Scotland to the Mediterranean.

Rissoidae

Rissoa rufilabrum (Fig. 52)

A species known mostly from western Scotland, where it occurs in a small number of brackish-water lochs and in fully marine conditions. Up to 5 mm high.

Rissoa parva (Fig. 52)

Occurs in some of the rocky brackish-water lochs of Scotland and elsewhere, but not widely recorded presumably because of the lack of appropriate hard and/or algal substrata.

Pusillina inconspicua (Fig. 52)

Occurs in a few of the rocky brackish-water lochs of Scotland, but rare elsewhere, presumably because of the lack of appropriate hard and/or algal substrata, since it can withstand salinities of down to 9‰. Up to 3 mm high.

Pusillina sarsi (Fig. 52)

Occurs quite commonly on *Zostera*, *Fucus* or even bare sediment in Danish brackish waters, but rarely recorded elsewhere, although it does occur from Norway to the Mediterranean. Sometimes recorded under the name *Rissoa albella*.

Cingula trifasciata (Fig. 52)

Occurs in some of the rocky brackish-water lochs of Scotland and in The Fleet, Dorset, but not elsewhere, presumably because of the lack of appropriate hard substrata.

Onoba aculeus (Fig. 52)

A small (<3 mm high), poorly known species that occurs on shingle in lagoons in England and amongst weeds in brackish-water lochs in Scotland. Southern England and Brittany are generally the southerly limit of its distribution, although there are two records from Spain.

Onoba semicostata (Fig. 52)

This species appears to be a southern version of *O. aculeus* above, occurring from Norway to the Mediterranean, although especially along western coasts. It is present in many Scottish brackish lochs, in Danish brackish waters and in some Channel lagoons. Up to 4 mm high.

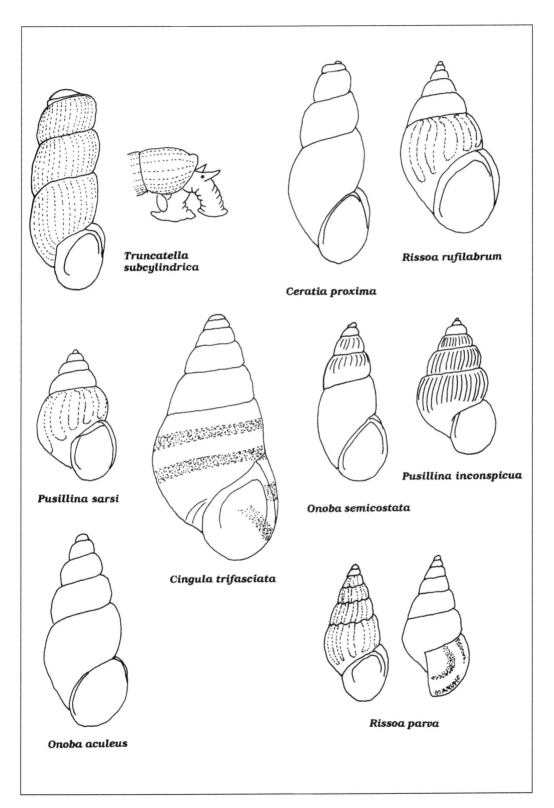

Truncatella
subcylindrica

Ceratia proxima

Rissoa rufilabrum

Pusillina sarsi

Cingula trifasciata

Onoba semicostata

Pusillina inconspicua

Onoba aculeus

Rissoa parva

Fig. 52. Brackish-water shelled gastropods (V).

Rissostomia membranacea/labiosa (Fig. 53)

A highly variable aggregate species that occurs down to 6‰ and is especially associated with meadows of *Zostera*. The two species *R. membranacea* and *R. labiosa* can be distinguished only by their method of juvenile development and by the form of the larval shell. Brackish-water populations of both produce few large eggs (>120 μm diam.) that hatch as baby snails, whilst in the sea many small eggs are laid (<100 μm diam.) each hatching into a veliger larva. The aggregate species occurs from Norway south to the Canaries, and in brackish water is commonest in the north (although it is abundant in The Fleet, Dorset, and in some Dutch lagoon-like ponds).

Assimineidae

Paludinella littorina (Fig. 53)

A small (up to 2 mm high) globose snail that occurs with *Truncatella* in lagoonal shingle in The Fleet, Dorset, and otherwise south to the Mediterranean in marine habitats.

Assiminea grayana (Fig. 53)

A locally common inhabitant of the terrestrial/brackish-water interface, especially the higher levels of salt-marshes and the margins of some lagoons, along a narrow band from eastern England to Denmark. Although the preferred habitat is rarely submerged, the eggs hatch into veliger larvae and must therefore await the highest spring tides to do so. Apparently, it can survive in fresh water for 2–3 days. Up to 5 mm tall.

Rissoellidae

Rissoella diaphana (Fig. 53); *Rissoella opalina* (Fig. 53); *Rissoella globularis* (Fig. 53)

These occur in several of the rocky brackish-water lochs of Scotland, and *R. diaphana* also occurs in The Fleet, but not elsewhere, presumably because of the lack of appropriate hard and/or algal substrata. All <2 mm high.

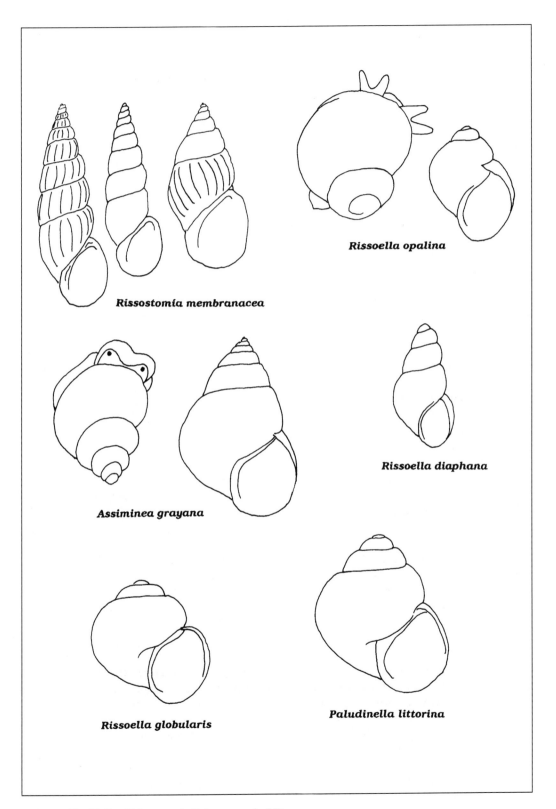

Fig. 53. Brackish-water shelled gastropods (VI).

Omalogyridae

Omalogyra atomus (Fig. 54)

Occurs on green algae such as *Ulva* and *Enteromorpha*, but under-recorded doubtless because of its minute size. Known from Norway to the Azores, including in lagoons. Up to 0.5 mm high and 1 mm diameter.

Skeneopsidae

Skeneopsis planorbis (Fig. 54)

Occurs in many of the rocky brackish-water lochs of Scotland, and on *Zostera* in Denmark, but rarely recorded elsewhere, presumably because like *Omalogyra* it is easily overlooked. Up to 0.8 mm high and 1.5 mm diameter.

Caecidae

Caecum armoricum (Fig. 54)

A minute snail (<1.5 mm long) restricted to Brittany and to The Fleet, Dorset. In the latter, it occurs in the shingle barrier, although little else is known of its lifestyle or habits (see Seaward, 1988).

Heteroglossa

Cerithiidae

Bittium reticulatum (Fig. 54)

A fairly common member of the *Ruppia* or *Zostera*-associated fauna, both on the plants and on the sediment beneath them, from Norway to the Black Sea, most often subtidally, but also present in a few lagoons outside the North Sea basin, where it may occur on shingle. Up to 8 mm high.

Stenoglossa

Muricidae

Nucella lapillus (dogwhelk) (Fig. 54)

The common dogwhelk penetrates into the more saline parts of estuaries, and the lagoons of the north, wherever its barnacle or mussel prey is also present. It is not, however, a normal member of the brackish-water fauna. Up to 30 mm tall.

Buccinidae

Buccinum undatum (whelk) (Fig. 54)

The whelk is essentially a subtidal marine species that is able to penetrate well into estuaries via the high-salinity water that is present at depth. From there, it occasionally moves up into the intertidal zone and is able to tolerate a considerable degree of brackishness (down to some 15‰).

Nassariidae

Hinia reticulata (Fig. 54)

Although *H. reticulata* (known in the older literature as *Nassarius*) is mainly a marine rocky-shore species, the variety *nitida* is characteristic of brackish waters down to nearly 15‰. This form is relatively small and has fewer costae and ridges than normal. It occurs in North Sea estuaries and in the western Baltic on *Zostera* and on most other substrata, including soft mud. Elsewhere *H. reticulata* occurs from Norway to the Mediterranean and Black Seas. Up to 25 mm high.

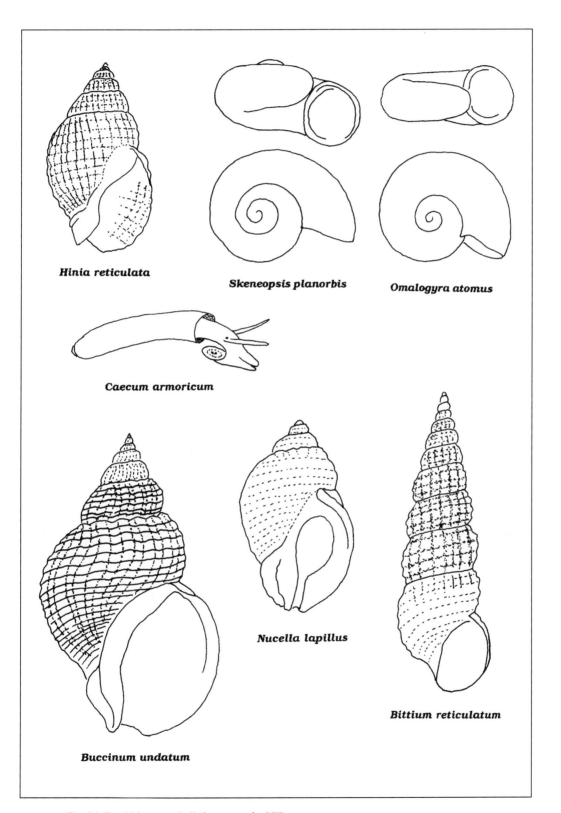

Hinia reticulata

Skeneopsis planorbis

Omalogyra atomus

Caecum armoricum

Nucella lapillus

Buccinum undatum

Bittium reticulatum

Fig. 54. Brackish-water shelled gastropods (VII).

GASTROPODA: ALLOGASTROPODA
Pyramidellomorpha
Pyramidellidae

Brachystomia eulimoides (Fig. 55)
A common micro-predator of scallops, *Turritella* and, in brackish waters, of oysters and occasionally mussels. Known from the Arctic to the Mediterranean but not recorded from the southern North Sea. <6 mm tall.

Brachystomia rissoides (Fig. 55)
A common micro-predator of the mussel *Mytilus* which occurs in northwestern Europe wherever its prey is found in waters of down to 10‰. Being only 3 mm tall or so, the pale shell is easily over-looked. In brackish waters it lays few large (100–120 μm diam.) eggs that hatch directly into baby snails, whereas in the sea it produces many small (55–70 μm diam.) eggs that hatch into planktonic larvae.

Turbonilla lactea (Fig. 55)
This small (<9 mm tall), white-shelled snail feeds on large, mud-dwelling polychaete worms (*Amphitrite, Cirriformia* and *Cirratulus*). Although apparently absent from the North Sea, it occurs from Norway to the Mediterranean.

GASTROPODA: OPISTHOBRANCHIA
Cephalaspida
Diaphanidae

Diaphana minuta (Fig. 55)
This small (<10 mm long), white, shelled species burrows in sand. Little is known of its biology. It has been recorded rarely from lagoon-like habitats in the north of the region. Otherwise known from the Arctic to the Canaries.

Retusidae

Retusa obtusa (Fig. 55)
A common, up to 15 mm long, whitish, shelled predator of, amongst others, forams and *Hydrobia*. Known from the North Atlantic and ?Pacific Oceans.

Retusa truncatula (Fig. 55)
A smaller (half size) version of *R. obtusa* with, apparently, similar biology and range.

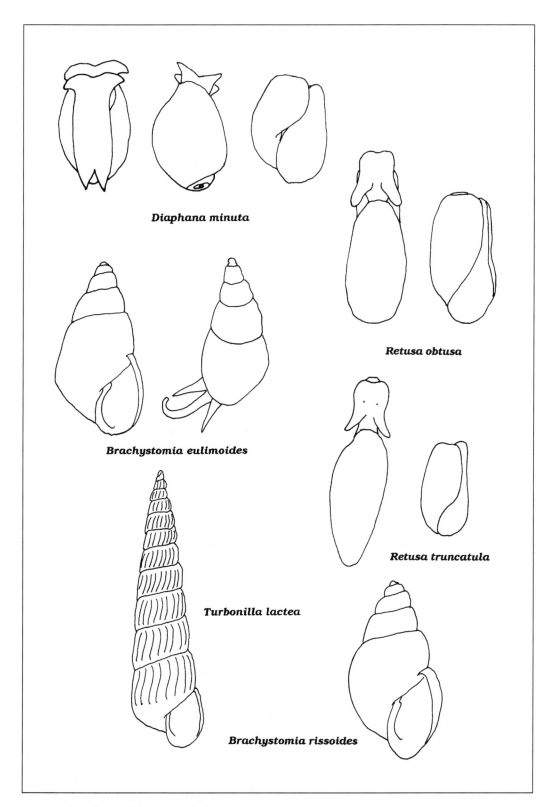

Fig. 55. Brackish-water shelled gastropods (VIII).

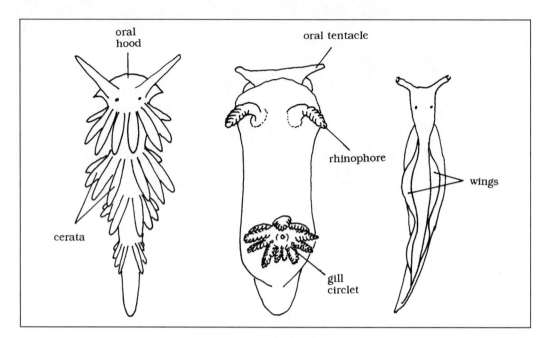

Fig. 56. Anatomical features used in the identification of shell-less gastropod molluscs.

Atyidae
Haminea navicula (Fig. 57)
A large (>50 mm long), but little-known species, particularly associated with *Zostera* meadows and recorded from Channel-coast lagoons (in Hampshire and Dorset). Widely distributed around the Atlantic.

Runcinidae
Runcina coronata (Fig. 57)
A small (<7 mm long) and little-known sea-slug recorded from green algae in lagoon-like habitats in Scotland and common in The Fleet. Elsewhere known from the European North Atlantic and Mediterranean.

Anaspida
Akeridae
Akera bullata (Fig. 57)
In The Fleet, Dorset, this herbivorous species occurs as the distinct variety *nana* : a small (shell <20 mm long), non-swimming form that cannot secrete purple dye. The more usual var. *farrani* (shell up to 40 mm long), which can also be found in lagoons and equivalent brackish waters, occurs from Norway to Spain and from Eire to Greece.

Pleurobranchomorpha
Pleurobranchidae
Berthella plumula (Fig. 57)
A large (up to 60 mm long), somewhat translucent, frothy- or lacy-looking, yellowish predator of the sea-squirt, *Botryllus*. Northeast Atlantic from Norway to Italy.

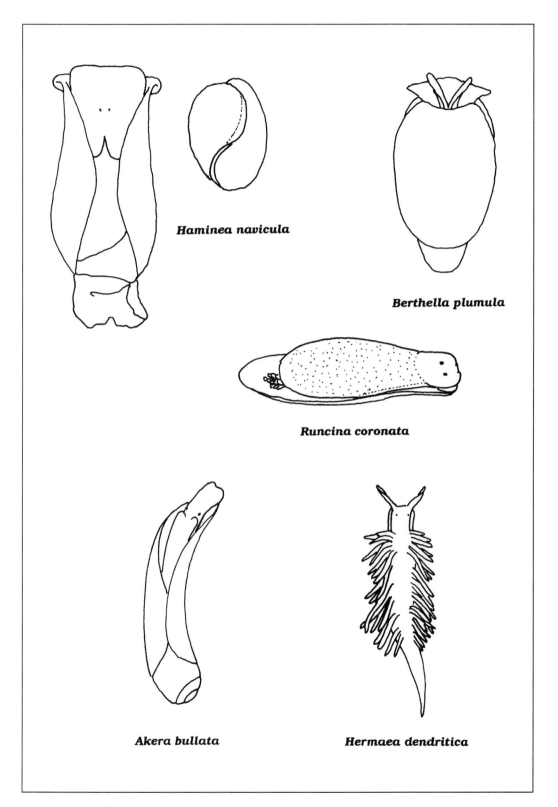

Fig. 57. Brackish-water sea-slugs (I).

Sacoglossa
Stiligeridae
Hermaea dendritica (Fig. 57)
A 10 mm long greenish consumer of green algae, known from Norway to Australia and from the Caribbean to Japan via the Mediterranean.

Stiliger bellulus (Fig. 58)
A rare, 10 mm long, grey or green (with black mottling and pale rhinophores) inhabitant of *Zostera* meadows. Known from Eire and Norway to the Mediterranean and Black Seas.

Alderia modesta (Fig. 58)
A widely distributed (Atlantic and Pacific) and common little sea-slug that is particularly associated with mats of *Vaucheria* on salt-marshes and sandy mudflats, and which can occur in salinities of down to 5‰. Up to 10 mm long.

Elysiidae
Elysia viridis (Fig. 58)
A fairly common, elongate, up to 45 mm long, consumer of algae, known from Norway to the Mediterranean.

Limapontiidae
Limapontia capitata (Fig. 58)
A small (usually <5 mm long) consumer of green algae (particularly *Chaetomorpha* and epiphytes of *Ruppia*) that has been recorded from some northern and eastern lagoon-like habitats and is otherwise known from the Arctic to the Mediterranean.

Limapontia depressa (Fig. 58)
A common, *c.* 6 mm long, consumer of *Vaucheria* on mudflats, especially on and near salt-marshes and in association with *Alderia* above, and of *Chaetomorpha* in tideless habitats. Known from the Baltic round to Atlantic France, in salinities of down to 4–6‰.

Limapontia senestra (Fig. 58)
Elsewhere known in fully marine rock pools from Norway to Atlantic France, but in rocky Scottish brackish-water lochs and in The Fleet, Dorset, associated with *Enteromorpha* and *Zostera* in salinities of down to some 20‰.

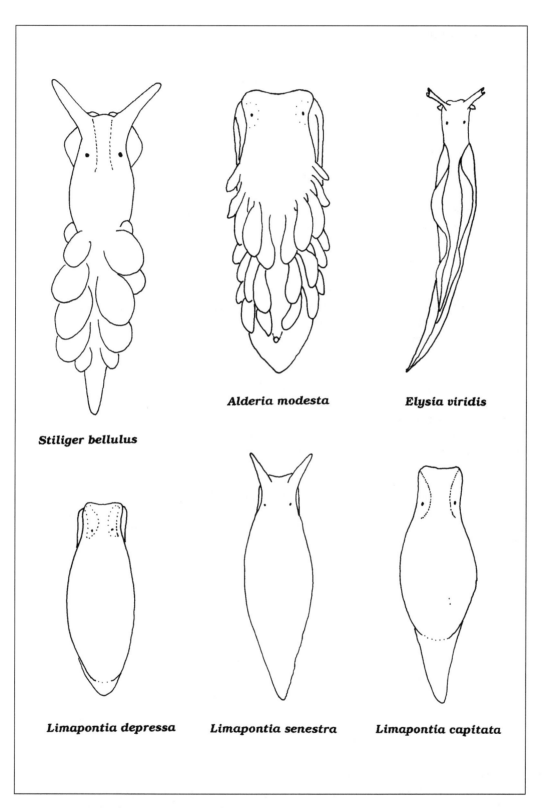

Stiliger bellulus

Alderia modesta

Elysia viridis

Limapontia depressa

Limapontia senestra

Limapontia capitata

Fig. 58. Brackish-water sea-slugs (II).

Nudibranchia
Dotidae
Doto coronata (Fig. 59)
A 10–15 mm long consumer of thecate hydroids (*Obelia, Dynamena* and *Sertularia* spp.) from the Arctic to the Mediterranean.

Corambidae
Doridella batava (Fig. 59)
A very small sea-slug (<5 mm long) known from the Zuiderzee and now feared extinct since the enclosure of that onetime arm of the Waddenzee. Its nearest relatives are to be found on the eastern seaboard of the U.S.A., and it is possible that this species was introduced from there into The Netherlands. Presumably, like other members of its genus it fed on encrusting bryozoans. Some records were from just outside the Zuiderzee and so it may survive.

Goniodorididae
Goniodoris nodosa (Fig. 59)
A common, up to 30 mm long consumer of encrusting bryozoans when young and sea-squirts (including *Botryllus*) when adult. Known from Norway to Spain in the northeast Atlantic.

Onchidorididae
Adalaria proxima (Fig. 59)
A pale, 20 mm long consumer of the bryozoan *Electra pilosa* in the northern half of the region. Widely distributed to the north of a line from Cape Cod to the Baltic.

Onchidoris muricata (Fig. 59)
Similarly to *Adalaria* above (with which it is easily confused), a consumer of *Electra pilosa* and similar species. Known from the northern regions of the North Atlantic and Pacific Oceans.

Acanthodoris pilosa (Fig. 59)
Like the other onchidorids, this species eats and is therefore associated with bryozoans. Its distribution is similar to that of *Onchidoris muricata*. Up to 55 mm long.

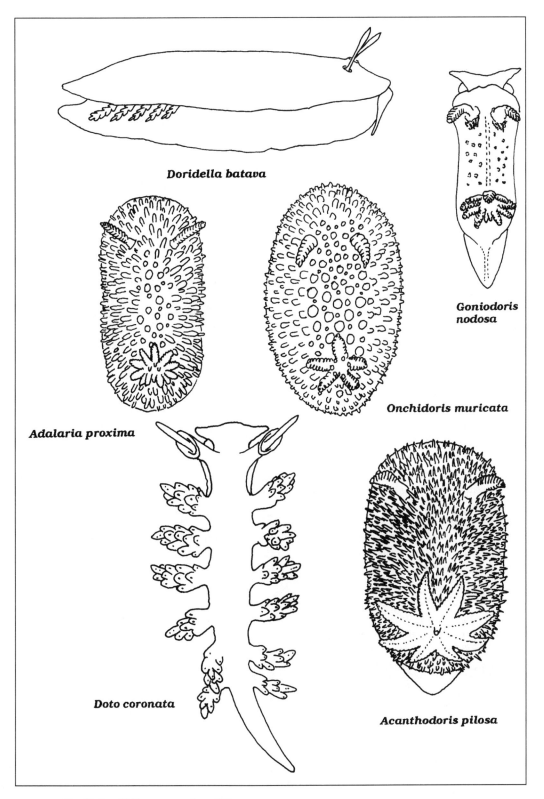

Fig. 59. Brackish-water sea-slugs (III).

Notodorididae

Aegires punctilucens (Fig. 60)

Recorded rarely from lagoon-like habitats in Scotland. Up to 20 mm long.

Cadlinidae

Cadlina laevis (Fig. 60)

Like *Aegires*, recorded rarely from Scotland, but unlikely normally to be a brackish-water species.

Archidorididae

Archidoris pseudoargus (Fig. 60)

A species that may just occur in the mouths of estuaries and estuarine lagoons, but not a regular member of the brackish-water fauna.

Tergipedidae

Tenellia adspersa (Fig. 60)

The most characteristically brackish-water nudibranch, occurring in salinities of down to 4‰. It feeds on hydroids (*Laomedea, Cordylophora, Protohydra,* etc.) in harbours, estuaries, lagoons, brackish ditches and canals. Egg number and size varies in different populations from many small (*c.* 70 μm diam.) to few large (140 μm diam.), and development correspondingly varies from being direct, through planktonic but non-feeding, to planktotrophic. In Denmark this variation correlates with salinity, but it does not do so in America. *Tenellia* is widespread (indeed is cosmopolitan), but in northwestern Europe it is somewhat sporadically distributed and seemingly is nowhere common. Being small (<10 mm long) and dully coloured (whitish, yellowish or brown with darker blotches) it can be easily overlooked.

Eubranchidae

Eubranchus farrani (Fig. 60)

Generally an inhabitant of kelp forests in which it consumes hydroids such as *Obelia*, but a small sea-slug (<10 mm long; half normal size) found in The Fleet, Dorset, in at least three colour varieties is assumed by Thompson (1988) to be a form or forms of this species.

Eubranchus exiguus (Fig. 60)

An Arctic and Baltic to Mediterranean and New England consumer of hydroids (*Obelia, Laomedea,* etc.) that can occur in salinities of down to 5‰. Like *Tenellia*, which it generally resembles in size and body colour though not in shape, it can be easily overlooked.

Aeolidiidae

Aeolidiella spp.

Although not specifically recorded from brackish habitats as far as the author is aware, it is likely that *Aeolidiella alderi, A. glauca* and *A. sanguinea*, which feed on the sea-anemones *Diadumene, Sagartia* and *Cereus*, will sooner or later turn up.

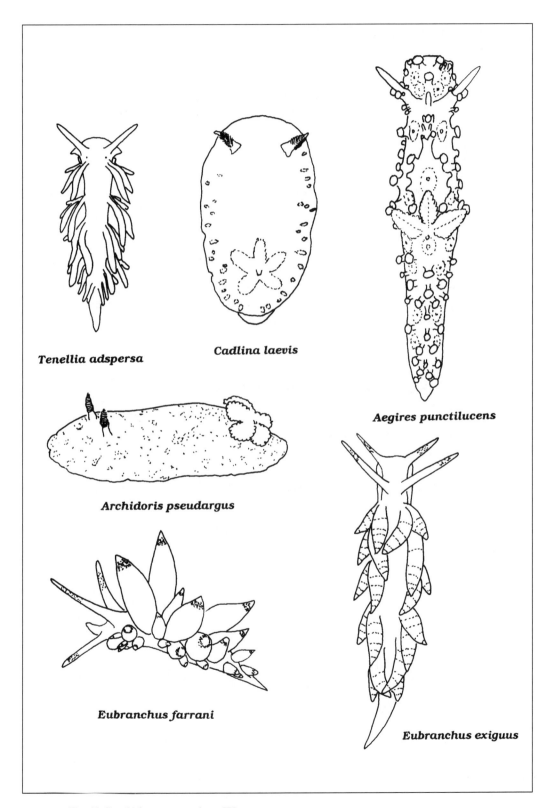

Tenellia adspersa

Cadlina laevis

Aegires punctilucens

Archidoris pseudargus

Eubranchus farrani

Eubranchus exiguus

Fig. 60. Brackish-water sea-slugs (IV).

GASTROPODA: PULMONATA
Archaeopulmonata
Ellobiidae

Ovatella myosotis (Fig. 62)
Often abundant beneath plant litter near the high-water mark in estuaries and salt-marshes, and just above the water level in lagoons. It appears able to survive over the entire brackish-water salinity range. Often recorded under the generic name *Phytia*. Up to 8 mm high.

Leucophytia bidentata (Fig. 62)
In similar habitats to *Ovatella* above, but also extending onto estuarine mudflats where it characteristically occurs in the cracks between the surface mud polygons that are caused by drying out at low tide. From southwestern shores of the North Sea westwards. Up to 6 mm high.

Basommatophora
Lymnaeidae

Lymnaea spp. (Fig. 62)
Pond-snails, especially *L. peregra* (<25 mm high), can extend out of their normal fresh water habitat and occur in the least saline brackish waters where these are stagnant or very slow flowing. Widely distributed throughout the north temperate zone of all continents.

Bivalves (cockles, mussels, clams, etc.)

Molluscs, ranging in shell length up to 20 cm, with the body wholly or partly enclosed within a hard, bivalved shell. Many species live buried in the sediments, but others, such as mussels, occur on the surface attached to stones and shells by byssus threads, and a few species are associated with submerged vegetation. Tebble (1966) describes British species. See Fig. 61.

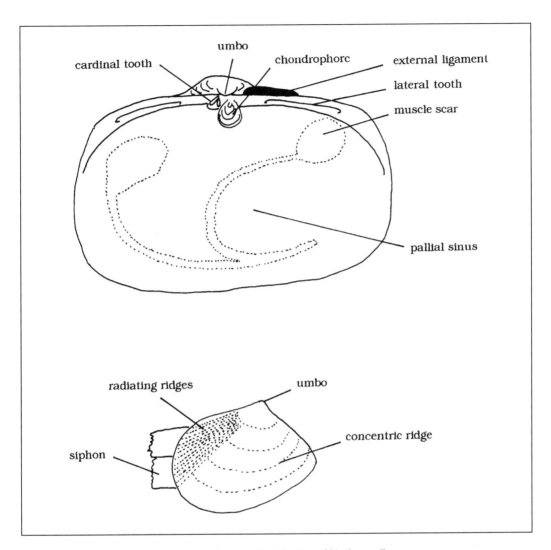

Fig. 61. Anatomical features used in the identification of bivalve molluscs.

n.b. This key, like the others, is designed to be used with living animals; on occasion, however, it may require investigation of the muscle and other tissue scars on the inside of the shells – empty shells usually abound where live animals are present, and after matching empty and live shells together, the empty ones can be used to supply information on these scars.

1   **A**   Animal unable completely to withdraw within its shell, the 2 siphons, bound together in a common sheath, remaining external to the shell at all times 2

     B   Animal capable of withdrawing entirely within its shell 3

2   **A**   Posterior margin of the shell (that from which the siphons issue) smoothly curved into a blunt, rounded point; large specimens >80 mm shell length .... *Mya arenaria*

B Posterior margin of shell abruptly truncate; shell <80 mm length...................... *Mya truncata*

3 A Shell subcircular, thin, brittle, flat, permanently attached via right valve to hard surface, including other bivalve shells 4
   B Shell not as above... 5

4 A Surface of shell (i.e. of left valve) with coarse concentric ridges and many radiating ridges; inner surface of left valve with 2 separate muscle scars; up to 45 mm diameter
   .... *Monia patelliformis*
   B Surface of left valve appearing smooth to the eye; inner surface of left valve with the 2 muscle scars touching; rarely up to 15 mm diameter........................... *Heteranomia squamula*

5 A Animal living on its left side on the surface of the substratum; valves dissimilar, the right (uppermost) ± flat, the left larger and domed 6
   B Either shell partially or wholly buried in the substratum, or, if exposed on the surface, then the two valves the same shape and size 7

6 A Shell subcircular; surface sculpture concentric, often forming flat scales (the native oyster)
   .... *Ostrea edulis*
   B Shell oval to very elongate; surface sculpture forming projecting frills, often with purple rays; umbones of left valve often curved over that of the right valve (the Pacific oyster)
   .... *Crassostrea gigas*

7 A Animal not buried or burrowing within the substatum, attached to hard objects, including other bivalve shells and vegetation, by a mat of fibres (the byssus)................................ 8
   B Animal buried or burrowing within the substratum; byssus, if present, not obvious............ 16

8 A Shell globular, with a surface sculpture of radiating ridges (ribs) over the whole surface of both valves; climbing amongst filamentous green algae or macrophytes........................... 9
   B Shell wedge-shaped, with umbones at the pointed (anterior) end; radiating ridges, if present, only over part of the valve surfaces; not climbing amongst macrophytes................... 11

9 A Right shell valve with 2 posterior lateral teeth; ligament small, arched, hidden behind umbones in lateral view; posterior dorsal margin not upcurved................ *Cerastoderma glaucum*
   B Right shell valve with 1 posterior lateral tooth; ligament long, flat, extending half way to posterior margin; posterior dorsal margin upcurved..10

10 A Shell oval in outline, with >22 ribs; right shell valve with 1 anterior lateral tooth
   .... *Parvicardium hauniense*
   B Shell almost triangular in outline, with 22 or less ribs; right shell valve with 2 anterior lateral teeth (the dorsal one often very small)....................................... *Parvicardium exiguum*

11 A Shell smooth, apart from growth lines, without sets of anterior and posterior radiating ridges.. 12
   B Shell with 2 sets of ridges radiating out from the umbones, a smaller anterior series and a more numerous posterior one.. 14

12 A Anterior margin extending slightly beyond the umbones; shell white-yellow, with red streaks.. *Modiolus adriaticus*

 B Umbones at extreme anterior end of the valves; shell not white or yellow, nor with red streaks.. 13

13 A Shell purple-blue, darker when larger, often translucent with blue and brown rays when young; anterior end without an internal platform; occurs over the marine half of the brackish-water spectrum .... *Mytilus edulis*

 B Shell greyish brown with white bands, zebra-striped when young; anterior end internally with a shelf or platform on which anterior adductor muscle inserts; occurs over the fresh water half of the brackish-water spectrum .... *Congeria cochleata*

14 A Umbones prominent; posterior-dorsal margin downcurved; with 15–18 ridges anteriorly .... *Modiolarca tumida*

 B Umbones not prominent; posterior-dorsal margin upcurved; with <15 anterior ridges.......... 15

15 A With >30 ridges in the posterior series; shell yellow-brown .... *Musculus discors*

 B With <30 ridges in the posterior series; shell whitish, often with red-brown chevrons along hinge line.. *Musculus costulatus*

16 A Shell minute (<3 mm long), fragile, plump, oval, white with tinges of red, occurring in shingle at the water level in lagoons.. *Lasaea rubra*

 B Shell not as above, occurring in habitats other than lagoonal shingle 17

17 A Shell surface covered by series of ridges (ribs) radiating from the umbones to the anterior, posterior and lower margins.. 18

 B Main surface sculpture of shell, if any, in form of concentric rather than radiating ridges and grooves 21

18 A Inside of shell valves with a series of furrows, each furrow corresponding to a rib on the outer surface, extending from shell margin to beneath umbones; shell valves often fragile and with the posterior region extended; characteristic of tideless waters 19

 B Furrows only extend a short distance up the inner shell surface, not approaching umbones; shell valves sturdy and with posterior region not extended; characteristic of tidal waters....... 20

19 A Right shell valve with 2 posterior lateral teeth; ligament small, arched and hidden by umbones in lateral view; posterior dorsal margin not upcurved; up to 50 mm long .... *Cerastoderma glaucum*

 B Right shell valve with 1 posterior lateral tooth; ligament long, flat, extending half way to posterior margin; posterior dorsal margin upcurved to give almost triangular outline to shell; <15 mm long... *Parvicardium exiguum* (lagoonal form)

20 A Right shell valve with 2 posterior lateral teeth; each valve with 22–28 ribs; up to 65 mm long (although usually much smaller in brackish waters) (the edible cockle)..... *Cerastoderma edule*

 B Right shell valve with 1 posterior lateral tooth; each valve with 20–22 ribs; <15 mm long .... *Parvicardium exiguum* (intertidal form)

21 A Shell elongate, up to 25 mm long, white to buff in colour, with some 10 concentric ridges raised up in the form of frills, particularly posteriorly .... *Irus irus*

B Shell without ridges raised up into projecting frills 22

22 A The 2 shell valves of markedly different sizes, the smaller left valve fitting into the larger right valve leaving a clear margin of the right valve uncovered; umbones ± central, ligament internal; shell white to cream, <15 mm long .... *Corbula gibba*

B Shell not as above 23

23 A Shell elongately rectangular, >4× as long as deep (6× in individuals >15 mm length), greyish violet in colour but covered by olive-green periostracum, with the umbones at one (anterior) end and with the two valves leaving an oval opening both anteriorly and posteriorly .... *Ensis directus*

B Shell <4× as long as deep 24

24 A Shell >2× as long as deep, white, with surface sculpture of concentric and radiating ridges drawn out into small spines at their intersections, <70 mm long; bores into peat, wood and soft rock .... *Barnea candida*

B Shell <2× as long as deep; not boring into soft rock, etc. 25

25 A Anterior dorsal region of shell with a well-marked heart-shaped area (the lunule) just in front of the umbones 26

B Shell without lunule 27

26 A Shell subtriangular, some 1.2× as long as deep; surface with concentric lines; inner margin of shell crenulate; inner surface with small V-shaped pallial sinus; length may exceed 12 cm .... *Mercenaria mercenaria*

B Shell oval, some 1.4× as long as deep; surface with concentric ridges and fine radiating lines, coarsest posteriorly; inner margin of shell smooth; inner surface with large U-shaped pallial sinus; <55 mm long .... *Venerupis senegalensis*
[Frequently recorded as V. pullastra*]*

27 A Shell very small (<10 mm long), fragile, white; inner surface without pallial sinus 28

B Shell not minute; the young, fragile stages of species with white shells all with a marked pallial sinus 29

28 A Shell almost circular, plump; umbones almost central; >3 mm long.............. *Kellia suborbicularis*

B Shell oval, with umbones clearly in posterior half; <3 mm long .... *Mysella bidentata*

29 A Outer surface covered by dark brown to black periostracum; inner surface without pallial sinus .... *Astarte borealis*

B Outer surface white to light brown (*beware* staining of *Scrobicularia* shells inhabiting black mud); inner surface with pallial sinus 30

30 A With short siphons bound together in a sheath; hinge with 2 of the cardinal teeth in left valve joined in form of inverted V; with small U-shaped pallial sinus; shell solid, subtriangular, dirty white to light brown .... *Spisula subtruncata*

B With elongate, separate, tubular siphons that can be extended out of the shell a distance >>shell length; hinge without cardinal teeth fused into an inverted V; pallial sinus large, extending at least half the distance between the muscle scars 31

31 A Shell plump, often pinkish or orange; umbones central, anterior half rounded, posterior region subtriangular; without internal ligament... *Macoma balthica*
   B Shell laterally flattened; with internal ligament in a chondrophore; never pinkish 32

32 A Shell circular or oval, greyish or fawn, with concentric lines and ridges, without lateral teeth in the right valve; up to 65 mm long... *Scrobicularia plana*
   B Shell white, without obvious surface sculpture; with 1 anterior and 1 posterior lateral tooth in right valve; <25 mm long... 33

33 A Umbones centrally placed; posterior margin truncate to smoothly rounded; shell dull white, length some 1.25×depth; <15 mm long.. *Abra tenuis*
   B Umbones in posterior half of shell; posterior margin bluntly pointed; shell glossy white, length some 1.5×depth; up to 25 mm long.. *Abra alba*

BIVALVIA
Mytilida
Mytilidae

Mytilus edulis (edible mussel) (Fig. 62)

The edible mussel is widely distributed through the northern hemisphere, including in estuaries down to salinities approaching 4‰, although growth rate is reduced in salinities of below 15–20‰. As a result of the ability of mussels to use empty shells as a suitable hard substratum, mussel beds can develop on relatively soft substrata: in quiet locations older generations of mussel shells become buried beneath a layer of pseudofaeces. Rarely present in lagoons. May occasionally exceed 200 mm length.

Modiolarca tumida (Fig. 62); *Musculus discors* (Fig. 62); *Musculus costulatus* (Fig. 62)

These three essentially rocky-shore species have been recorded from rocky lagoon-like habitats in Scotland, and in Denmark at least *M. discors* occurs on *Zostera*; their absence alsewhere presumably reflects the lack of suitable substrata to the south. Length <20 mm.

Modiolus adriaticus (Fig. 62)

The variety *ovalis* occurs in estuaries in southwestern England, otherwise this is an offshore species found from the Baltic to the Mediterranean and Black Seas. In the Kattegat it is associated with *Zostera*. Length <60 mm.

Pteriida
Ostreidae

Crassostrea gigas (Japanese or Pacific oyster) (Fig. 62)

Layings of this introduced oyster have been made in the mouths of a number of estuaries; however it breeds only sporadically. Up to 200 mm long.

Ostrea edulis (native oyster) (Fig. 62)

Wild populations of this oyster are now rare, but cultivated and managed stocks occur in several estuaries in salinities of down to 20‰. Usually <120 mm diameter.

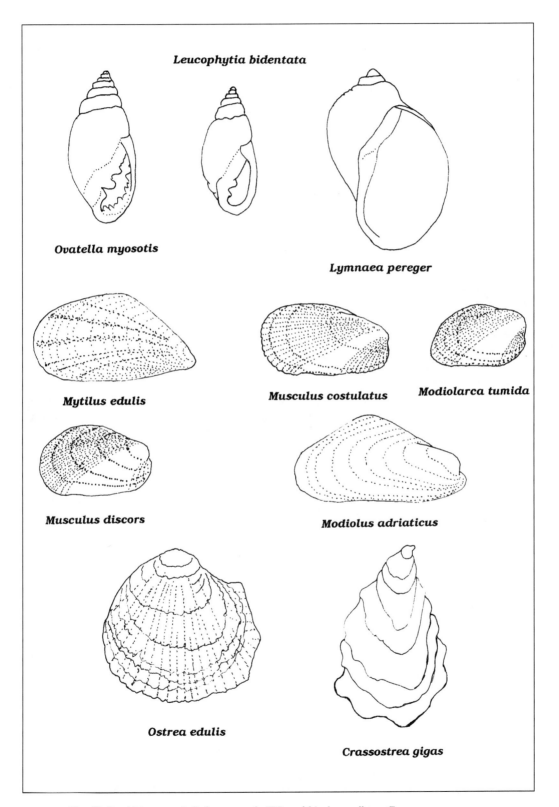

Fig. 62. Brackish-water shelled gastropods (IX) and bivalve molluscs (I).

Anomiidae (saddle-oysters)

Heteranomia squamula (Fig. 63); *Monia patelliformis* (Fig. 63)

These two essentially rocky-shore species have been recorded from rocky lagoon-like habitats in Scotland; their absence elsewhere presumably reflects the lack of suitable substrata to the south.

Venerida

Astartidae

Astarte borealis (Fig. 63)

A northern species that occurs (in relatively deep waters) in Denmark, and can withstand salinities of down to 8‰. <50 mm long.

Kellidae

Kellia suborbicularis (Fig. 63)

Recorded from silty shingle and stones from Scotland to The Fleet. Virtually a cosmopolitan species. Up to 10 mm long.

Lasaeidae

Lasaea rubra (Fig. 63)

A member of the water-level shingle fauna in The Fleet, Dorset; otherwise a widespread crevice- or algal-associated species from Norway to the Canaries.

Montacutidae

Mysella bidentata (Fig. 63)

A small species that in brackish waters may be commensal in the burrows of *Nereis, Barnea* and others, or may occur free in soft sediments of both estuaries and lagoons. It has been recorded as part of the associated fauna of *Chaetomorpha*, attached by its byssal threads, and in shingle. Elsewhere, it occurs from Norway to the equator, and in the Mediterranean and Black Seas.

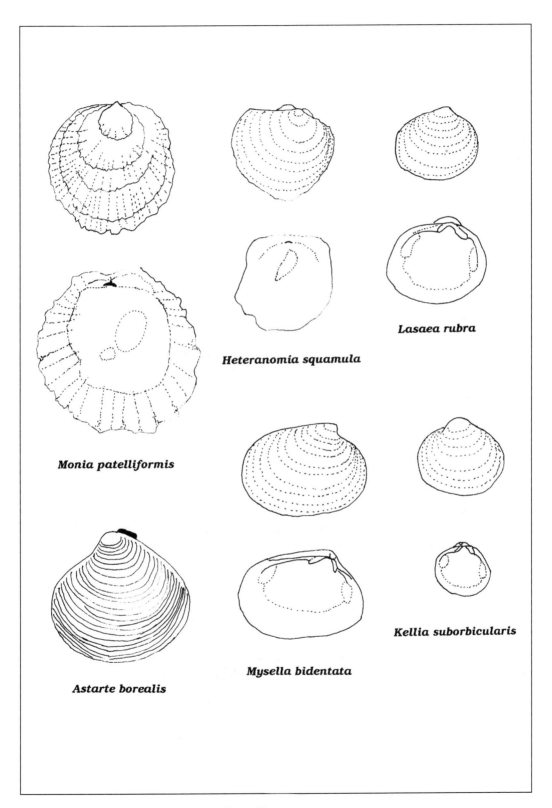

Fig. 63. Brackish-water bivalve molluscs (II).

Cardiidae

Cerastoderma glaucum (lagoon cockle) (Fig. 64)

A widespread and common lagoonal species, often recorded under the name *Cardium lamarcki*, that replaces *C. edule* in such habitats. It can occur down to 4‰ salinity. The young stages (and occasionally the adults) live attached to vegetation such as *Chaetomorpha*, *Ruppia* and, more rarely, *Zostera* or *Potamogeton*; the adults usually burrow only shallowly even in soft substrata. Occurs from Norway and the Baltic to the Mediterranean and Black Seas, although because of past confusion with *C. edule* its distribution is imperfectly known. Empty shells are a favoured site for egg laying in the goby *Pomatoschistus microps*.

Cerastoderma edule (edible cockle) (Fig. 64)

Replaces *C. glaucum* in estuaries in which it can occur, burrowing in soft sediments, down to some 15‰, although in brackish waters it usually achieves a much smaller size than in the sea. Distributed from Norway to Morocco. Records from the Mediterranean to the Caspian Seas are erroneus.

Parvicardium exiguum (Fig. 64)

Occurs in all types of brackish water down to some 20‰, although a different form occurs in each of tidal and non-tidal conditions (varying in shell form, including the teeth, in periostracum, etc.; see Petersen & Russell, 1971b). Like *Cerastoderma glaucum*, the lagoonal form occurs in submerged vegetation such as *Enteromorpha* and charophytes; in contrast, the estuarine form lives on the surface of the sediment. Known from Norway to the Black Sea but is largely replaced by *P. hauniense* in the Kattegat and Baltic.

Parvicardium hauniense (Fig. 64)

A Baltic species that extends westwards as far as the waddens, estuaries and lagoons of Denmark in areas with salinities of down to 6‰; it has not been recorded from salinities of >12‰. *P. hauniense* occurs climbing in submerged vegetation (*Zostera, Ruppia,* charophytes and, occasionally, fucoid algae) by means of its byssus, it has not been found in or on the bottom sediments. <10 mm long.

Veneridae

Venerupis senegalensis (Fig. 64)

Widely recorded as *V. pullastra*, this is a common and widespread species in gravelly and other coarse sediments from Norway to Morocco, burrowing a few centimetres below the surface. It occurs in the mouths of most estuaries where firm, coarser-than-mud substrata are present, but does not penetrate below some 20‰.

Mercenaria mercenaria (hard-shelled clam) (Fig. 64)

An American species introduced to Europe in the middle of the nineteenth century and now found in scattered estuaries, and a few estuarine lagoons, throughout the region. A naturalized population is especially abundant in and around Southampton Water and the Seudre Estuary in Poitou-Charentes, France.

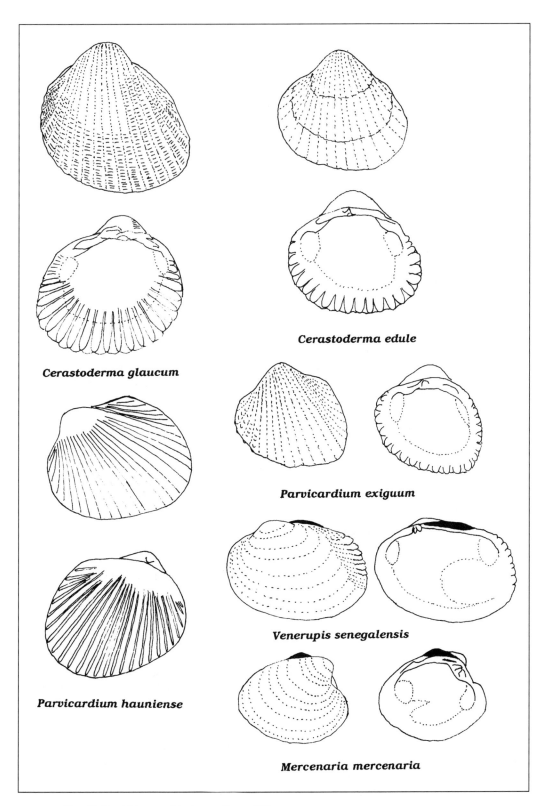

Cerastoderma edule

Cerastoderma glaucum

Parvicardium exiguum

Parvicardium hauniense

Venerupis senegalensis

Mercenaria mercenaria

Fig. 64. Brackish-water bivalve molluscs (III).

Irus irus (Fig. 65)

Recorded once from the bed of The Fleet, Dorset, and rather more frequently from within shingle at the water level there, but not otherwise a brackish-water species.

Mactridae

Spisula subtruncata (Fig. 65)

A common subtidal bivalve in the mouths of estuaries in relatively sandy sediments where the salinity is in excess of 15‰. It is distributed from Norway to the Canaries and in the Mediterranean and Black Seas. <30 mm long.

Tellinidae

Macoma balthica (Fig. 65)

One of the most characteristic bivalves of estuaries and also present in many lagoons, in salinities of down to 5‰. It is a mobile, relatively shallowly-burrowing species with elongate siphons, the inhalent member of which can draw a water column from the overlying water or from the sediment/water interface. Densities can exceed 5000 individuals/m². Like the scrobiculariids (below) it lies on its side within the sediment. <30 mm long.

Scrobiculariidae

Abra tenuis (Fig. 65)

An abundant small (<15 mm long), but surprisingly little known, inhabitant of lagoonal and intertidal estuarine muds. It appears to be a southerly species, occurring from north Africa and the Mediterranean northwards to western Scotland and northern England. In tideless water it can occur in salinities down to 10‰.

Abra alba (Fig. 65)

Like *A. tenuis* a burrower in estuarine muds, but commoner in deeper waters and less frequent than that species in lagoons. It occurs from Norway and the Baltic to the Mediterranean and Black Seas and West Africa, in salinities of down to near 10‰ in the Baltic, but higher than this (*c.* 20-25‰) elsewhere.

Scrobicularia plana (Fig. 65)

A characteristic and widely distributed deep-burrower in estuarine muds that can occur in salinities of down to 10‰ (and exceptionally 4‰) and down to depths below the surface of some 30 cm. Feeding is effected by means of a long, mobile inhalent siphon that vacuum-cleans the surface sediment around the burrow, often leaving a pattern of radiating spokes. It may attain an age approaching 20 years and densities of over 1000 individuals/m². Although occurring northwards to Norway and in the Baltic, it is a southerly species and fared badly in the 1962–63 cold winter.

Solenidae

Ensis directus (the American jack-knife clam) (Fig. 65)

This razor-shell is a recent introduction from the eastern coast of North America (first recorded in 1979) that now occurs all along the southern shore of the North Sea (i.e. from France to Denmark). It may occur, burrowing in sandy sediments, in the mouths of estuaries in this region, but usually only subtidally. Exceptionally up to 250 mm long in America, but thus far much less in Europe.

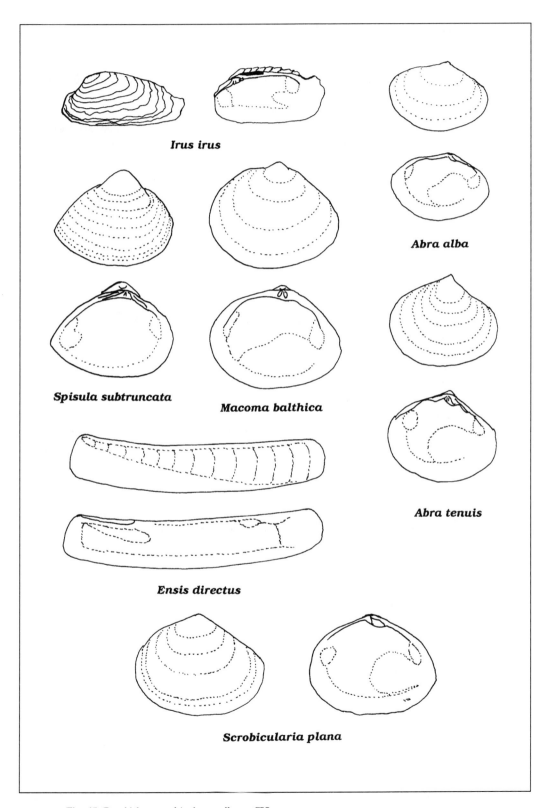

Fig. 65. Brackish-water bivalve molluscs (IV).

Dreissenidae
Congeria cochleata (Fig. 66)
This species was introduced into Belgium from the subtropical Caribbean in the early years of the 19th Century. It has since spread along the southern coast of the North Sea into France, The Netherlands and Germany where it lives attached to hard surfaces by a byssus in low-salinity ditches, canals and lagoon-like habitats, and may there cause fouling problems to, for example, power stations. In spite of belonging to a largely fresh water family, a planktonic larval stage has been retained. It has recently been suggested that its correct name is *Mytilopsis leucophaeta*. <50 mm long.

Corbulidae
Corbula gibba (Fig. 66)
Widely distributed in relatively deep water around the mouths of estuaries in salinities in excess of 15‰.

Myida
Myacidae
Mya arenaria (soft-shelled clam or sand-gaper) (Fig. 66)
A characteristic, suspension-feeding, deep-burrower of brackish-water sediments that occurs right across the northern part of the northern hemisphere in salinities of down to 4‰, although active life may require salinities in excess of 10–15‰. The siphons are permanently extended and, with care, their double-tubed tips can be observed flush with the sediment surface: the body itself may be up to 0.5 m below. Lifespan is up to some 20 years.

Mya truncata (blunt-gaper) (Fig. 66)
In many respects a very similar species to *M. arenaria* (above) and with a similar sediment and salinity range and life style; it is less common intertidally, however, and is essentially a subtidal estuarine/marine species. In the Baltic it occurs down to 10‰ salinity.

Pholadidae
Barnea candida (Fig. 66)
Burrows in outcrops of peat and in stiff clay in estuaries down to salinities of some 20‰.

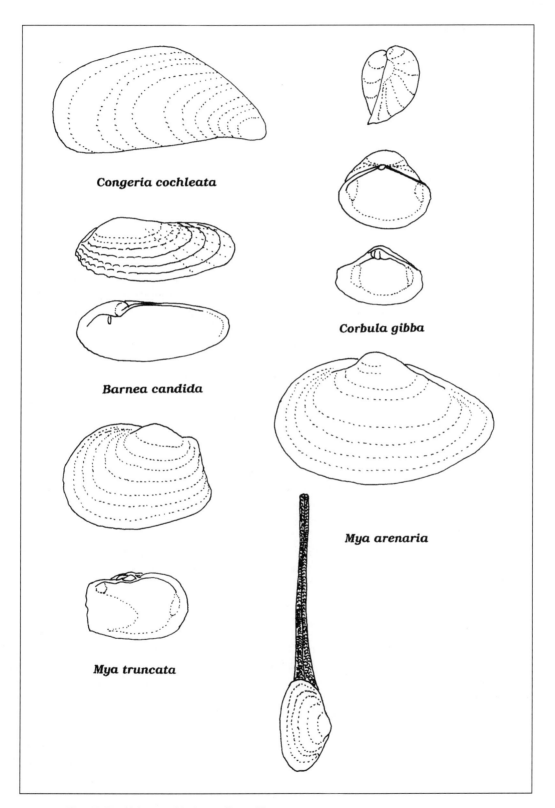

Fig. 66. Brackish-water bivalve molluscs (V).

Crustaceans

Barnacles

Small (<16 mm diameter), squatly conical, circular-in-outline, whitish animals firmly cemented to stones and other hard objects. The body is hidden beneath 4-6 large, radially arranged calcareous plates and an operculum of 4 small plates at the apex of the cone. Bassindale (1964) describes British species.

1 A Shell with 4 radial plates .... *Elminius modestus*
  B Shell with 6 radial plates 2

2 A Radial shell plates smooth, white, with or without coloured stripes; base calcareous but porous and thin (bases of dead and eroded individuals will be evident as ± circular, calcareous films on the substratum, scattered amongst the living individuals).............................. 3
  B Radial shell plates ridged; base membraneous...................................... *Semibalanus balanoides*

3 A Radial shell plates without pink, purple or brown stripes; not associated with areas warmed by power-station discharges... *Balanus improvisus*
  B Radial shell plates with groups of pink, purple or brown stripes; restricted to areas warmed by power-station discharges.. *Balanus amphitrite*

Thoracica
Balanidae
Balanus improvisus (Fig. 67)
An introduced barnacle characteristic of brackish waters but capable of inhabiting both sea and almost fresh waters, in respect of the latter down to below 3‰ in the Finnish Baltic. It is especially common on hard substrata, algae, etc. in estuaries. Up to 15 mm diameter

Balanus amphitrite (Fig. 67)
An introduced, warm-water species (originally from the Red Sea) that is mainly restricted to areas, including estuarine docks and harbours, warmed by cooling water discharges from power stations. Up to 15 mm diameter.

Semibalanus balanoides (Fig. 67)
Essentially a marine species that can withstand some dilution of its environment, down to 20‰ or so, and can therefore colonize hard substrata in the mouths of estuaries. It is placed in the genus *Balanus* in the older literature. Up to 15 mm diameter.

Elminius modestus (Fig. 67)
An introduced Antipodean species that was first recorded in the mid 1940s and is still spreading. Its current local distribution is centred in the west but includes the southern North Sea and most major ports. It appears to be displacing *Balanus improvisus* where the two come into contact. Up to 10 mm diameter.

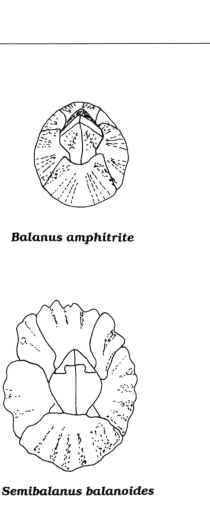

Balanus amphitrite

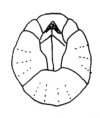

Balanus improvisus

Semibalanus balanoides

Elminius modestus

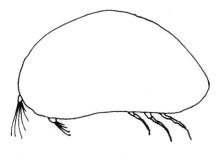

ostracod

Fig. 67. Brackish-water barnacles and ostracods.

Ostracods

Small (<1 mm long) crustaceans with the body enclosed within a bean-shaped, bivalved shell, from which protrude a few pairs of appendages during movement (Fig. 67). May be abundant in lagoonal waters and surface sediments. See Athersuch *et al.* (1989).

Malacostracans

The following six groups are all malacostracan crustaceans, and not surprisingly therefore they share a number of features. Their bodies comprise a head, an 8-segmented thorax, a 6-segmented abdomen, and a terminal telson, although various of these segments may be fused together and/or hidden beneath an anterior carapace (the whole of a crab's body is hidden by the carapace when viewed from above). Their appendages are differentiated into: 2 pairs of anterior antennae (one of which may be much longer than the other), various mouthparts, relatively stout walking legs (pereiopods) which may or may not terminate in chelae or subchelae, abdominal swimming legs (pleopods) and 'tail legs' (uropods).

Mysids (oppossum shrimps)

Semi-transparent, cylindrical, swimming crustaceans resembling small (<26 mm long) delicate shrimps. A carapace covers the head and (partially) the thorax, and possesses a characteristic groove across it, dividing it into two sections. The pereiopods are obviously biramous. Makings (1977) describes British species. See Fig. 68.

1 A Telson not terminally cleft 2
  B Telson terminally bifid.. 5

2 A Telson short, <2× as long as broad... *Mesopodopsis slabberi*
  B Telson elongate, >3× as long as broad 3

3 A Inner branch of uropod without any spines along the distal half of its inner margin
    .... *Neomysis integer*
  B Inner branch of uropod with spines along the whole length of its inner margin................... 4

4 A Body surface (including of the appendages) transparent but covered in fine scales; spines on inner margin of inner branch of uropod irregular in size and not increasing uniformly in size distally .... *Leptomysis gracilis*
  B Body surface appearing opaque and brown, without scales; spines on inner margin of inner branch of uropod regularly increasing in size distally .... *Leptomysis mediterranea*

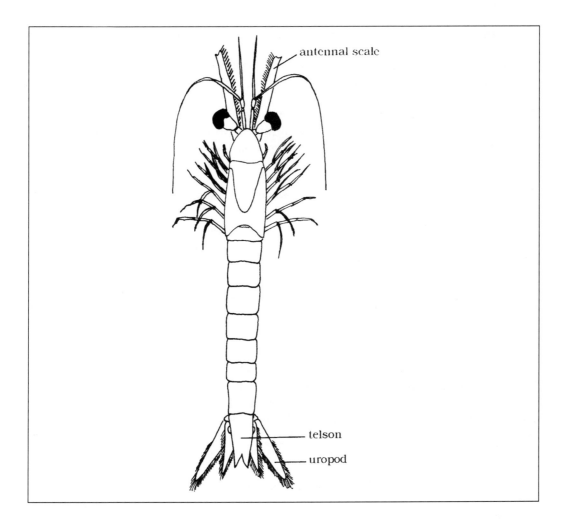

Fig. 68. Anatomical features used in the identification of mysid crustaceans.

5 A Penultimate segment with a backwardly projecting mid-dorsal spine on its posterior border
.... *Gastrosaccus spinifer*

B Penultimate segment without a mid-dorsal spine 6

6 A Outer margin of outer branch of uropod with series of long spines increasing in size distally .... *Gastrosaccus sanctus*

B Outer branch of uropod without spines on its outer margin 7

7 A Inner branch of uropod curved inwards along its distal third, its inner margin with three very long spines at the base of the curved portion and one such at its tip ...... *Schistomysis parkeri*

B Inner branch of uropod not as described above 8

8 A Sides of telson each with <20 spines 9

B Sides of telson each with >20 spines 10

9 **A** Inner margin of inner branch of uropod with 5 or 6 small spines proximally; antennal scale 4× as long as broad .... *Praunus inermis*

  B Inner margin of inner branch of uropod with >10 spines distributed over most of its length; antennal scale 3× as long as broad.. *Paramysis nouveli*

10 **A** Antennal scale elongate and parallel-sided, 7–8× as long as broad, with spine at its distal end .... *Praunus flexuosus*

  B Antennal scale oval, <6× as long as broad, with spine not at its tip but 60–70 % of the distance along the margin.. 11

11 **A** Spines on the inner margin of the inner branch of the uropod extending almost to the distal end .... *Schistomysis ornata*

  B Distal fifth of the inner margin of the inner branch of the uropod without spines 12

12 **A** Animal colourless, transparent; eyes elongate, extending well beyond the sides of the carapace .... *Schistomysis spiritus*

  B Animal greyish white, with spots of yellow, red or brown; eyes short not or scarcely extending beyond the sides of the carapace .... *Schistomysis kervillei*

Mysidae

Neomysis integer (Fig. 69)

This is one of the two or three most characteristic brackish-water mysids. *N. integer* (often termed *N. vulgaris* in the early and in the continental literature) is a dominant and almost ubiquitous inhabitant of water columns in the range 0.5–20‰ salinity (and more rarely in adjacent fresh waters and in waters of greater than 20‰) both in estuaries and in non-tidal lagoons and ponds. Shoals of >4000 individuals/m^2 and feeding rates of up to 6.5×10^6 diatom cells per mysid per hour have been recorded. It occurs throughout the region, but has very rarely been observed in the sea; it appears to have adapted successfully to the transition from brackish lagoon to fresh water lagoon in some areas (e.g. the Loch Mor Barvas lagoon on Lewis). Up to 20 mm long.

Praunus flexuosus (Fig. 69)

Although *P. flexuosus* can occur down to some 6‰ salinity and rarely lower, it tends to replace *Neomysis* in estuaries and lagoons throughout the region over the range 20–35‰ It is notable for the ability to change its colour from being effectively transparent to being almost black, and for probably being the region's most abundant mysid; it is not known south of Roscoff, Brittany. Unlike many mysids, it is not associated with the bottom but has been described as spending much time 'hovering' head uppermost in the water. Up to 25 mm long.

Praunus inermis (Fig. 69)

Like its congener above, *P. inermis* occurs northwards from Roscoff; it has occasionally been recorded from amongst *Ulva* and other seaweeds in the mouths of estuaries. It occurs in low-salinity waters in the Baltic. Up to 20 mm long.

Paramysis nouveli (Fig. 69)

A relatively rare, shallow-water species that has been recorded from high-salinity lagoons in East Anglia, and from estuaries in southwestern England and from the Clyde. It is essentially a Mediterranean species. Up to 12 mm long.

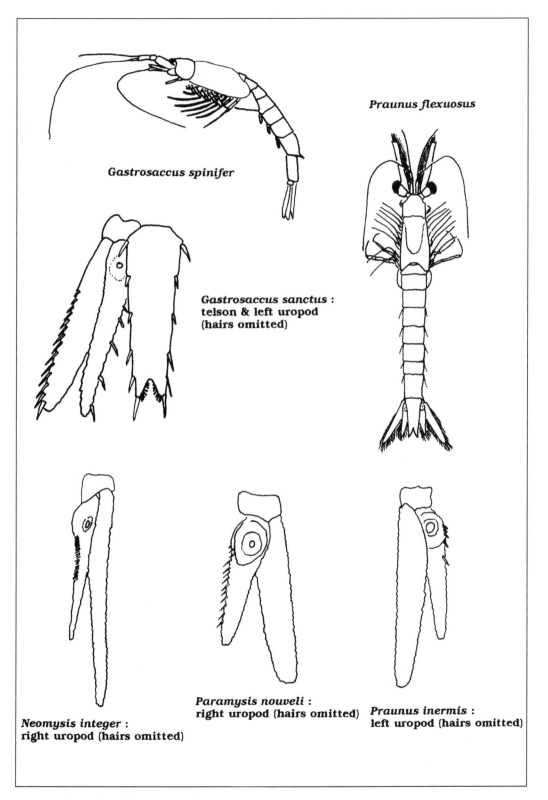

Gastrosaccus spinifer

Praunus flexuosus

Gastrosaccus sanctus :
telson & left uropod
(hairs omitted)

Neomysis integer :
right uropod (hairs omitted)

Paramysis nouveli :
right uropod (hairs omitted)

Praunus inermis :
left uropod (hairs omitted)

Fig. 69. Brackish-water mysids (I).

Gastrosaccus spinifer (Fig. 69)
A widely distributed and common, bottom-dwelling species that can extend into estuaries to about 18‰ salinity (and in the Baltic into more dilute waters). It occurs in dense swarms, including in *Zostera* meadows, and is often found partially buried in the sediment. Up to 20 mm long.

Gastrosaccus sanctus (Fig. 69)
An uncommon southwestern species that extends into the region as far as the southern shores of the North Sea and that can occur in the mouths of estuaries, in waters of down to some 18‰. Up to 15 mm long.

Mesopodopsis slabberi (Fig. 70)
An abundant and characteristic estuarine mysid throughout the region, occurring down to 0.5‰ salinity and able to withstand fresh water for short periods. It has been recorded from virtually all northwest European estuaries, sometimes in shoals so numerous that the water has been described as 'gelatinous' as a result. Apart from the eyes, however, it is completely transparent, and hence often only visible after being caught in a net. Up to 15 mm long.

Leptomysis gracilis (Fig. 70)
Essentially a coastal northeast Atlantic species that may occasionally penetrate into the mouths of estuaries in the region. Up to 15 mm long.

Leptomysis mediterranea (Fig. 70)
Essentially a Mediterranean species that extends as far as the southwest of the region, and occasionally into the North Sea. It has been recorded from the mouths of some English estuaries. Up to 20 mm long.

Schistomysis kervillei (Fig. 70)
A relatively rare species known from the southern shore of the Channel and North Sea northwards to Scotland, including in the mouths of the Mersey and Seine Estuaries. Rarely over 20 mm long.

Schistomysis ornata (Fig. 70)
A widespread species occurring throughout the region although mainly in fully marine coastal waters. It has been recorded from some estuaries (e.g. of the Thames, Severn, Mersey and Seine) down to some 20‰, but it does not seem often to occur in brackish waters. Up to 20 mm long.

Schistomysis parkeri (Fig. 70)
A southern species that just reaches southwest England and that may occur in the mouths of estuaries there (e.g. the Exe Estuary). Rarely >10 mm long.

Schistomysis spiritus (Fig. 70)
A common coastal species throughout the region that sometimes penetrates into the mouths of estuaries. Up to 20 mm long.

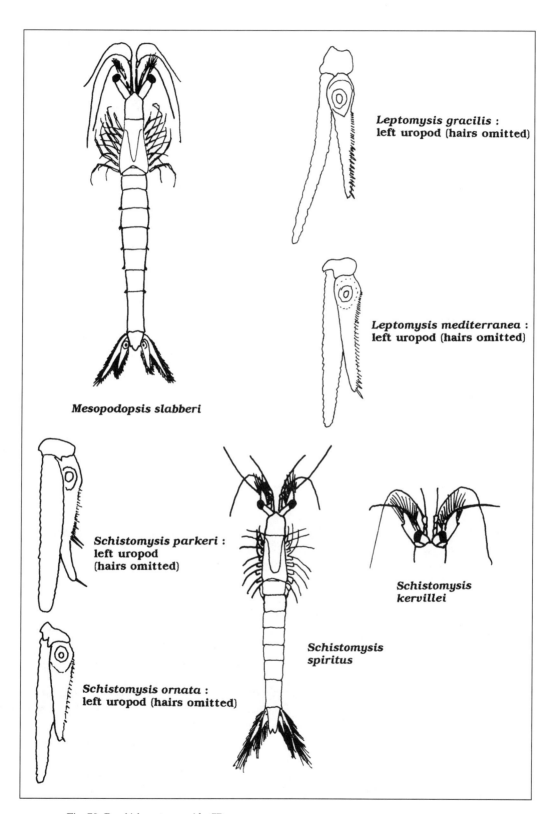

Leptomysis gracilis :
left uropod (hairs omitted)

Leptomysis mediterranea :
left uropod (hairs omitted)

Mesopodopsis slabberi

Schistomysis parkeri :
left uropod
(hairs omitted)

*Schistomysis
kervillei*

*Schistomysis
spiritus*

Schistomysis ornata :
left uropod (hairs omitted)

Fig. 70. Brackish-water mysids (II).

Cumaceans

The only cumacean likely to be encountered is *Diastylis rathkei*. This burrows shallowly into soft sediments leaving the anterior carapace and uropods exposed, and may also swim up into the water column. Jones (1976) describes the British species.

Diastylidae
Diastylis rathkei (Fig. 72)
A northern species that occurs in the surface layers of mud, or occasionally coarser sediments, in some estuaries, for example the Severn, and also in the Baltic, besides more widely in the marine environment. Up to 20 mm long.

Isopods (woodlouse-like animals)

Small aquatic woodlice (<30 mm long) that are flattened when viewed from above, do not have any of their segments covered by a carapace, and possess 5–7 pairs of virtually identical uniramous walking legs, and a single pair of uropods. Occur on the sediment surface (or more rarely in burrows in it) or beneath stones or swim amongst submerged vegetation. Naylor (1972) and Gledhill *et al.* (1976) describe British species. See Fig. 71.

1 A With 5 pairs of walking legs; adult male with large head and projecting jaws; adult female and juveniles with swollen thorax.. *Paragnathia formica*
  B With 7 pairs of walking legs; adult male without large projecting jaws; adult female without swollen thorax 2

2 A Body about 10× longer than wide .... *Cyathura carinata*
  B Body <5× longer than wide 3

3 A Body comprising besides a head and a 7-segmented thorax, a single, undivided abdominal plate 4
  B Body comprising besides a head and a 7-segmented thorax, an abdomen of more than one apparent segment.. 6

4 A Uropods elongate, projecting beyond end of abdomen; body not scale-like, >3× longer than broad; up to 10 mm length .... *Asellus aquaticus*
  B Uropods minute, recessed into an abdominal notch; body scale-like, <2.5× longer than greatest breadth, <5 mm in length 5

5 A Body oval, densely fringed by spines; male praeoperculum (the 1st pleopod modified for copulation) narrow, pointed .... *Jaera nordmanni*
  B Body straight-sided, sparsely fringed by spines; male praeoperculum T-shaped .... *Jaera albifrons* agg. (See separate key below)

6 A Abdomen of 6 apparent segments 7
  B Abdomen with <6 apparent segments as a result of partial fusion.................................... 8

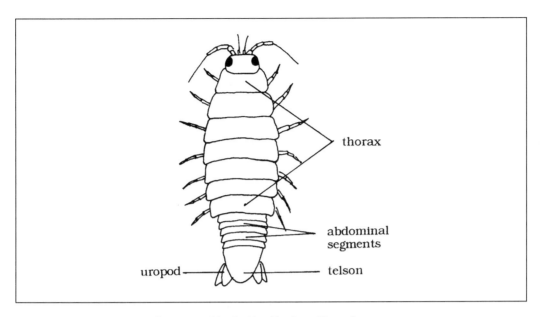

Fig. 71. Anatomical features used in the identification of isopod crustaceans.

7 A 1st abdominal segment distinctly narrower than last thoracic one when viewed from above, giving the animal a tadpole-shape; <8 mm in length; occurs on sand .... *Eurydice pulchra*

 B Abdomen not sharply set off from the thorax, woodlouse-shaped; up to 30 mm in length; occurs on rocks and other hard substrata .... *Ligia oceanica*

8 A Animal oval, <2.5× as long as broad; uropods evident, mounted on side of last apparent body segment 9

 B Animal elongate, >3× longer than broad; uropods concealed beneath last apparent body segment.. 11

9 A Last body segment with 2 longitudinal rows of tall tubercles on its dorsal surface
 .... *Sphaeroma hookeri*

 B Last body segment smooth or with scattered granules but without tubercles arranged in 2 rows 10

10 A Last body segment with scattered granules; outer margin of outer branch of uropod smooth .... *Sphaeroma rugicauda*

 B Last body segment smooth; outer margin of outer branch of uropod slightly serrated
 .... *Sphaeroma monodi*

11 A Tip of last body segment tridentate; body up to 30 mm length .... *Idotea balthica*

 B Tip of last body segment with single point; body length <20 mm 12

12 A Sides of last body segment slightly concave, its tip with a distinct acute point; body broadest in middle of thorax; posterior thoracic segments with pointed posterior angles; body some 3–4× longer than maximum breadth... *Idotea granulosa*

B Sides of last body segment straight, its tip with an indistinct point; body broadest across the last thoracic segments, which have rounded posterior angles; body 4–5× longer than broad (except in egg-carrying females)... *Idotea chelipes*

Key to members of the *Jaera albifrons* group
n.b. only the males of these minute isopods can be identified to species.

1 A Propodus, carpus and merus of 1st–4th pairs of legs with dense curved setae (carpus, for example, with >10 such setae).. *J. praehirsuta*
  B Propodus, carpus and merus of 1st–4th pairs of legs without dense curved setae (carpus, for example, with <6 such setae) 2

2 A Ischium of 6th and 7th pairs of legs with dense cluster of curved setae................ *J. ischiosetosa*
  B Ischium of 6th and 7th pairs of legs without dense cluster of curved setae........................ 3

3 A Distal region of carpus of 6th and 7th pairs of legs expanded into a lobe fringed by up to 40 spines .... *J. albifrons*
  B Distal region of carpus of 6th and 7th pairs of legs with only 2–4 spines and not expanded into a lobe.. *J. forsmani*

Gnathiidae
Paragnathia formica (Fig. 72)
A characteristically estuarine and salt-marsh species that typically occurs near the top of the sides of salt-marsh creeks (at the high-water level of neap tides) in burrows in each of which a male guards a hareem of up to 20 females. The praniza larvae leave the burrows during flooding tides and attach to fish such as flounders or gobies, consuming their blood and body fluids. Salinities down to some 18‰ can be tolerated. Up to 5 mm long.

Anthuridae
Cyathura carinata (Fig. 72)
This inhabitant of vertical tubes in stable, sandy or muddy sediments in estuaries, and more rarely lagoons, occurs somewhat sporadically throughout the region in salinities of down to less than 1‰. In many localities, only the occasional individual can be found, yet densities of 3000 individuals/m^2 are known at other sites. Its range extends from the Baltic to Ireland and to the Mediterranean, but is present only over the southern portion of the region. Up to 15 mm long.

Cirolanidae
Euridice pulchra (Fig. 72)
An inhabitant of intertidal sands throughout the region that can occupy such sediments in the mouths of estuaries down to some 18‰ salinity.

Sphaeromatidae
Sphaeroma hookeri (Fig. 72)
Although it may occur at high tidal levels at the head of sheltered estuaries, this species is most often found permanently submerged on the sediment or amongst submerged vegetation (especially *Ruppia*) in low-salinity lagoons and creeks (salinity 1–10‰, and occasionally higher, up to 35‰) in which it replaces other *Sphaeroma* species. It occurs throughout the region. It has recently (1987) been transferred to the genus *Lekanesphaera*. Up to 11 mm long.

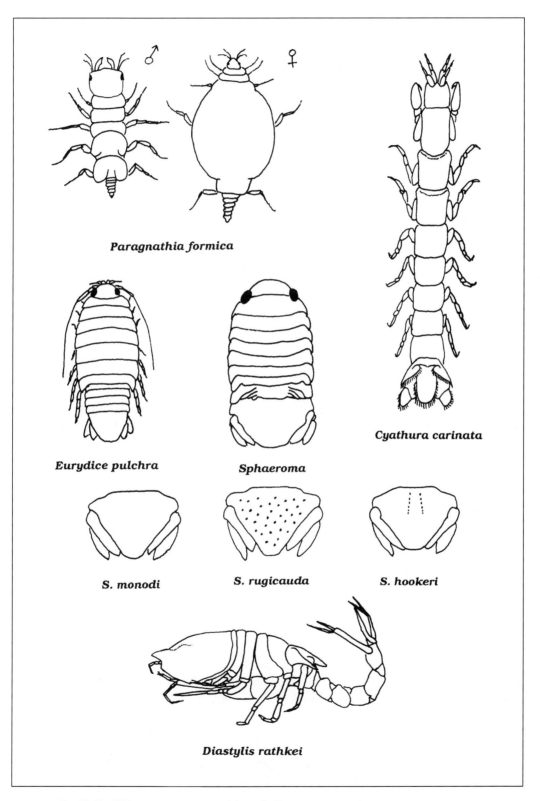

Paragnathia formica

Cyathura carinata

Eurydice pulchra

Sphaeroma

S. monodi

S. rugicauda

S. hookeri

Diastylis rathkei

Fig. 72. Brackish-water cumacean and isopods (I).

Sphaeroma rugicauda (Fig. 72)

A characteristic estuarine salt-marsh species that occurs in salt pans and in relatively high-salinity lagoons in salinities of over 8‰, although rarely it may be found down to 4‰. It typically occurs in vegetation (in salt pans, especially the overhanging vegetation around the margin), but it may also occupy under-stone and crevice habitats. It has been recorded throughout the region. It has recently (1987) been transferred to the genus *Lekanesphaera*. Up to 10 mm long.

Sphaeroma monodi (Fig. 72)

A southern species, penetrating the region as far as the southwestern shores of the North Sea, that occurs under stones over the lower half of the estuarine intertidal zone, most commonly in salinities of more than 14‰, and also where fresh water discharges across marine beaches. It has recently (1987) been transferred to the genus *Lekanesphaera* under the name *L. levii*. Up to 12 mm long.

Idoteidae

Idotea chelipes (Fig. 73)

The main habitat of *I. chelipes*, which was known for many years under the name *I. viridis*, is amongst submerged vegetation (*Ruppia, Chaetomorpha*, etc.) in non-tidal lagoons, brackish ponds and ditches in salinities of down to 4‰, although in the Baltic it extends down to 2‰, and it has also been recorded from amongst algae in highly sheltered estuaries. It occurs throughout the region. Up to 15 mm long.

Idotea balthica (Fig. 73) and *Idotea granulosa* (Fig. 73)

These two species replace *I. chelipes* in areas of greater water movement. In the Baltic, they can extend down to 6–8‰, although in other northwest European brackish waters they appear to inhabit higher salinity regions, often of more than 18‰. Both species are estuarine, occurring amongst fucoid and other macroalgae which they consume when it starts to decay. *I. granulosa* can attain 20 mm length, *I. balthica* half as long again.

Asellidae

Asellus aquaticus (Fig. 73)

This essentially fresh water species only occasionally penetrates brackish-waters, usually in lagoons and similar non-tidal habitats, up to salinities of the order of 6‰, and rarely of more than 8‰.

Janiridae

Jaera nordmanni (Fig. 73)

A minute isopod occurring beneath stones in low salinity lagoons and over the upper half of the beach where fresh water streams discharge down the shore, especially in the south and west of the region. Usually found in salinities of <5‰, but in one lagoonal record (Channel coast of England) the salinity was 22‰. <5 mm long.

Jaera albifrons agg. (Fig. 73)

A complex of four species distinguishable only in the male sex. The aggregate *J. albifrons* is termed *J. marina* in the older literature. The four species, which penetrate brackish water to varying degrees are:

J. ischiosetosa – the most characteristically brackish-water of the four, occurring throughout the region in lagoons (for example on *Ruppia*) and beneath stones in areas of fresh water discharge on the shore, down to some 2‰. <5 mm long.

J. albifrons (sensu stricto) – occurring beneath stones, etc. in shallow pools at high tidal levels on estuarine shores throughout northwest Europe. <5 mm long.

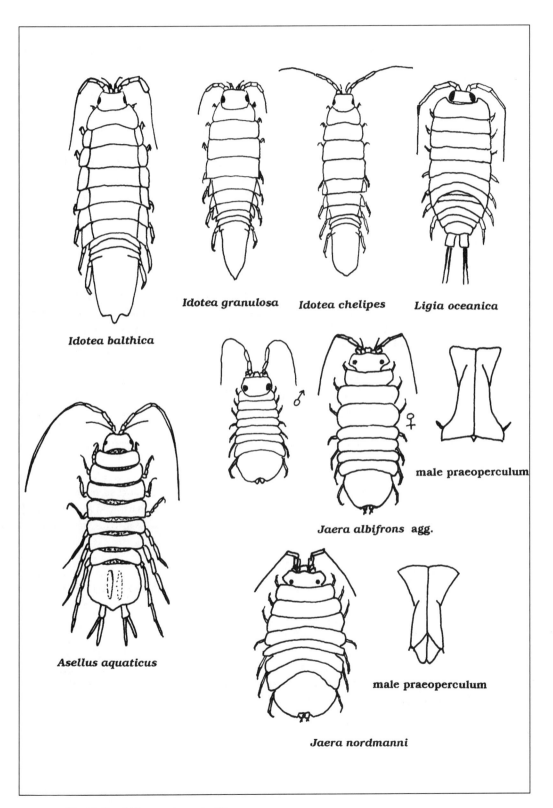

Fig. 73. Brackish-water isopods (II).

J. praehirsuta – occurring beneath fucoid algae at high tidal levels on the shores of estuaries and highly sheltered marine inlets throughout the region; tolerance of 2‰ has been recorded. <5 mm long.

J. forsmani – the least euryhaline of the four, it is a southwestern species occurring beneath damp stones at high tidal levels in the mouths of some estuaries. Up to 6 mm long.

Ligiidae
Ligia oceanica (Fig. 73)
A large, air-breathing woodlouse that occurs in crevices on hard substrata at high tidal levels throughout the coastal zone, including in estuaries and just above the water level of lagoons. It can often be observed running over rocks, stones and concrete surfaces.

Tanaids

Small (<10 mm long), more or less cylindrical or dorsoventrally flattened, whitish crustaceans with a carapace covering the head and first two segments (leaving 12 or less apparent segments visible), and with large, chela-bearing, first pereiopods, followed by 6 virtually identical uniramous walking legs. Live in tubes or burrows in the sediment or on vegetation and other structures. Holdich & Jones (1983) describe British species.

1 A First 2 segments of abdomen each with a semicircular row of strong erect setae across the dorsal surface .... *Tanais dulongii*
  B Abdominal segments without any dorsal rows of setae 2

2 A First walking leg larger and stouter than the others (used for digging); sides of abdominal segments produced into spines when viewed from above; cephalothorax not sexually dimorphic .... *Apseudes latreillii*
  B First pair of walking legs similar to the other 5 pairs; sides of abdominal segments not produced into spines; cephalothorax sexually dimorphic, that of males narrowed and elongated anteriorly .... *Heterotanais oerstedi*

Apseudidae
Apseudes latreillii (Fig. 75)
Sporadically recorded as occurring in muddy gravel and *Zostera* sediments from eastern Scotland to France, including (rarely) in brackish regions. Body length <8 mm.

Tanaidae
Tanais dulongii (Fig. 75)
A widespread species that inhabits tubes of sediment and detritus, and that occasionally penetrates into brackish water, occurring, for example, in several Channel coast lagoons with salinities of >20‰. Up to 5 mm long.

Paratanaidae

Heterotanais oerstedi (Fig. 75)

Occurs throughout the region in small U-shaped tubes attached to hydroids, algae or other submerged vegetation and to hard surfaces in estuarine and non-tidal brackish habitats down to some 4‰ salinity, including being present in the Baltic. Its biology and range are little known, doubtless because of its minute size, although it does genuinely appear to be patchy in its distribution. It can, however, occur in densities of over 5000 individuals/m². Males seem rare, and the species may be a protandrous hermaphrodite. <3 mm long.

Amphipods (sandhoppers, etc.)

Small crustaceans (<35 mm long) that are flattened from side to side, do not have any of their segments covered by a carapace, and possess 7 pairs of markedly heterogeneous uniramous pereiopods, in particular the first 2 pairs usually being subchelate, and 3 pairs of uropods. Often a somewhat translucent greyish white in colour. In burrows in the sediment, in tubes on vegetation, beneath stones and algae, amongst submerged plants, etc. Lincoln (1979) and Gledhill *et al.* (1976) describe British species.

n.b. This key uses a number of abbreviations: A = antenna; G = gnathopod; P = pereiopod; and U = uropod. Numbers after these letters refer to the precise appendage concerned; e.g. G2 is the second gnathopod, and U3 is the third uropod. See Fig. 74 for explanation of these and other terms.

1 A Wood-boring animal; pleosome segment 3 with a long curved process dorsally; U3 extremely large with a flat, spiny outer branch.. *Chelura terebrans*
   B Non-wood-boring species; pleosome segment 3 and U3 not as above 2

2 A Urosome segments with groups of spines and setae on the dorsal surface
          .... see separate key to Gammarids below
   B Urosome segments with no more than 1 pair of short dorsal spines and/or setae each, although the posterior dorsal margins may bear backwardly-directed teeth and the first urosome segment may have one or more keels or other projecting structures in the midline 3

3 A A2 large, robust, with 7 apparent articles or less, prehensile or raptorial, without an evident flagellum.. see separate key to *Corophium* below
   B A2 with at least 8 articles, including a flagellum.. 4

4 A Pleosome segments with backwardly-projecting dorsal 'beaks'........................ *Dexamine spinosa*
   B Pleosome segments without such beaks 5

5 A Pleosome and posterior pereon segments with median, dorsal, toothed keel, as on a crocodile's tail.. *Gammarellus angulosus*
   B Body without such a keel 6

6 A Dorsal surface of 1st urosome segment with some form of keel or other process or processes in the midline.. 7
   B Dorsal surface of 1st urosome segment without projections 9

7 A Dorsal surface of 1st urosome segment with a single, forked projection .... *Pontoporeia femorata*
   B Projection on the 1st urosome segment not a forked process 8

8 A Dorsal surface of 1st urosome segment with a single keel down the midline
          .... *Gammarella fucicola*
   B Dorsal surface of 1st urosome segment with 2 keel-like processes, one behind the other
          .... *Atylus swammerdami*

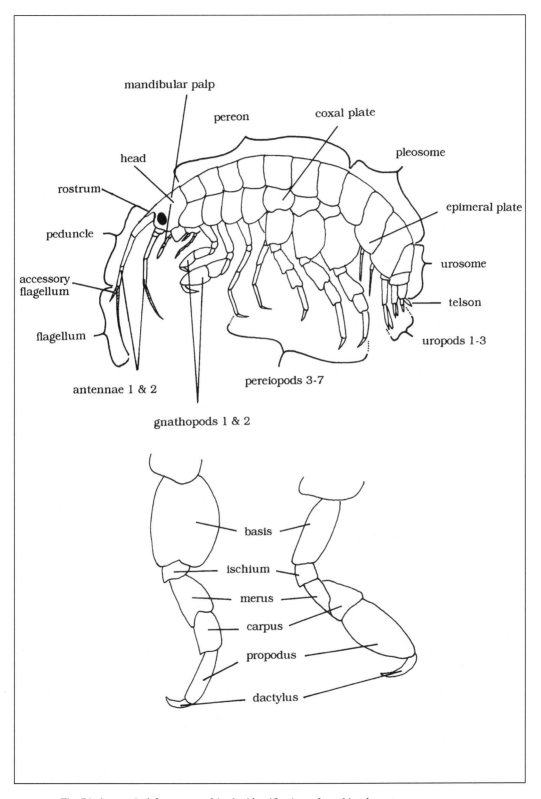

Fig. 74. Anatomical features used in the identification of amphipod crustaceans.

9 A Anterior margin of basis of G2 with forwardly-directed fringe of very long, dense setae .... *Leptocheirus pilosus*
  B Without such setae on basis of G2.. 10

10 A G1 chelate by virtue of carpal process which extends distally to tip of propodus (dactylus in usual subchelate form) .... *Leucothoe* spp.
  B G1 not as above 11

11 A Some pereiopods clearly enlarged and flattened for digging 12
  B Pereiopods not obviously adapted for digging 13

12 A Meri and carpi of P6 and P7 expanded... *Haustorius arenarius*
  B Meri and carpi of P6 and P7 not expanded; but meri, carpi and propodi of P5 expanded and spinose.. *Urothoe* spp.

13 A First article of peduncle of A1 forming a 'roman nose', rest of A1 issuing from its lower surface.. *Bathyporeia pilosa*
  B Peduncle of A1 not as above.. 14

14 A G2 slender, elongate, with terminal articles not enlarged and not functionally subchelate; ischium of G2 elongate, longer than merus; carpus and propodus of G2 setose; A1 peduncle large and broad .... *Lysianassa ceratina*
  B G2 and A1 not as above 15

15 A Squat amphipods with coxal plates 2–4 large and shield-like 16
  B Coxal plates 2–4 not especially enlarged and shield-like 18

16 A U2 projecting much less than U1 or U3; 4th coxal plate emarginate posteriorly .... *Gitana sarsi*
  B U2 not projecting much less than U1 or U3; 4th coxal plate not emarginate posteriorly 17

17 A Telson with 2–3 pairs of dorsolateral spines; body greenish white with brown patches .... *Metopa pusilla*
  B Telson without any dorsolateral spines; body whitish with red markings ..... *Stenothoe monoculoides*

18 A Head with two pairs anterior eyes; A1 small; base of A2 concealed beneath coxal plate; basis of P7 with expanded posterior lobe.. *Ampelisca brevicornis*
  B Without this combination of features.. 19

19 A G2 long, slender and chelate; P7 twice length of other pereiopods; eyes contiguous dorsally, at base of rostrum .... *Pontocrates* spp.
  B Without this combination of features.. 20

20 A P6 much longer than other pereiopods; head with rostral hood covering bases of the short antennae.. *Phoxocephalus holbolli*
  B Without this combination of features.. 21

21 A A1 distinctly shorter than A2 22

B A1 subequal to or longer than A2 28

22 A Urosome segments 1 and 2 with their posterior dorsal margin drawn out into 3–4 backwardly-projecting median teeth .... *Cheirocratus sundevalli*

B Urosome segments 1 and 2 without such teeth 23

23 A With prominent accessory flagellum on A1, equal in length to some 50% of that of the main flagellum .... *Liljeborgia kinahani*

B A1 accessory flagellum, if present, small, <25% of the length of main flagellum 24

24 A Flagellum of A2 short, with only 5 articles 25

B Flagellum of A2 long, with 10 or more articles 26

25 A Propodus of G2 either with large forwardly-directed process arising proximally on lower margin (in males) or with lower margin concave between rounded proximal lobe and angular distal process (in females); without dense setae along lower margin .... *Jassa falcata*

B Propodus of G2 either with a densely setose straight or slightly concave lower margin and lacking a forwardly-directed process arising proximally (in males) or lower margin sinuous but generally convex and without any concavity between two processes (females) .... *Ischyrocerus anguipes*

26 A Tip of A1 extending beyond the peduncle of A2 .... *Hyale nilssoni*

B A1 very short, shorter than the A2 peduncle 27

27 A U3 ending in large spine almost equal in length to the distal article; G2 never large and subchelate; G1 larger and more robust than G2 in both sexes .... *Talitrus saltator*

B U3 not terminating in a spine almost as long as the distal article; G2 large and sub-chelate in males; G1 smaller than G2 in both sexes .... *Orchestia* spp.

28 A Telson laminar and elongate; 4th coxal plate emarginate posteriorly 29

B Telson short and fleshy; 4th coxal plate not emarginate posteriorly 33

29 A A1 without accessory flagellum 30

B A1 with accessory flagellum 31

30 A Last peduncular article of A1 with forwardly projecting ventral point .... *Calliopius laeviusculus*

B Last peduncular article of A1 without such a projection .... *Apherusa* spp.

31 A The two branches of U3 subequal .... *Maera grossimana*

B Inner branch of U3 <outer branch 32

32 A Posterior dorsal margin of urosome segment 1 with a backwardly-projecting median tooth .... *Melita palmata*

B Posterior dorsal margin of urosome segments without teeth .... *Melita pellucida*

33 A The two branches of U3 shorter than the basal peduncle 34

B The two branches of U3 equal to or greater than length of basal peduncle 36

34 A 3rd article of peduncle of A1 <half length of 2nd article .... *Ampithoe rubricata*

B 3rd article of peduncle of A1 >half length of 2nd article 35

35 A U3 uniramous .... *Ericthonius* spp.

B U3 biramous .... *Ischyrocerus anguipes*

36 A A1 >>A2 in length; G1 larger than G2 37

B A1 and A2 subequal; G2 larger than G1 39

37 Only the males of the 3 spp. of aorids that key out here can be distinguished; aorid females are often remarkably similar to each other. Males can be distinguished by their large 2nd gnathopods.

A Merus of male G1 with very large, long, pointed process extending as far as the propodus
.... *Aora typica*

B Merus of male G1 without any such process, although other articles may bear processes 38

38 A Carpus of male G1 with 1+ tooth-like distal processes; upper margins of carpus and propodus of male G1 without dense mats of setae .... *Microdeutopus gryllotalpa*

B Carpus of male G1 without such processes, although process present on propodus; upper margins of carpus and propodus of male G1 with dense mats of setae .... *Lembos websteri*

39 A Base of A2 hidden beneath coxal plate; antennae without marked hair fringes; coxal plates with fringe of hairs; small (<3 mm); body dark brown .... *Microprotopus maculatus*

B Base of A2 exposed; antennae with distinct fringes of long hairs; coxal plates with only a few setae and no fringe; body pale yellow .... *Gammaropsis maculata*

Key to Gammarids

1 A Inner branch of U3 small, less than 0.2 length of outer branch 2

B Inner branch of U3 larger, at least 0.3 length of outer branch 6

2 A Telson lobes each with 1 apical spine; body white (characteristic of clean shingle)
.... *Pectenogammarus planicrurus*

B Telson lobes each with 3 apical spines; body dark, mainly various shades of green 3

3 A G2 propodus smaller than that of G1; G2 carpus elongate, >2× as long as deep
.... *Eulimnogammarus obtusatus*

B G2 propodus larger than that of G1; G2 carpus short and triangular, <2× as long as deep 4

4 A A2 and outer branch of U3 with dense marginal setae 5

B A2 and outer branch of U3 without dense marginal setae .... *Chaetogammarus stoerensis*

5 A Dorsal surface of urosome segments each with >15 spines forming transverse rows; length of merus of P7 3× width.. *Chaetogammarus marinus*

  B Urosome with <10 dorsal spines on each segment; merus of P7 with length 2× width .... *Chaetogammarus pirloti*

6 A Eyes small, almost circular, their length not more than twice their width (a fresh water species only occasionally straying into brackish conditions) .... *Gammarus pulex*

  B Eyes large, kidney-shaped, their length more than twice their width 7

7 A Inner and outer branches of U3 subequal.. 8

  B Inner branch of U3 <90% length of outer... 10

8 A Basis of P7 only slightly longer than wide; urosome segments without marked dorsal humps .... *Gammarus crinicornis*

  B Length of basis of P7 at least 1.5× width; urosome segments with marked dorsal humps 9

9 A Posterior margin of 3rd epimeral plate with many small setae; length of basis of P7 some 1.5× width.. *Gammarus locusta*

  B Posterior margin of 3rd epimeral plate with 0 or 1 small seta; length of basis of P7 >1.5× width... *Gammarus insensibilis*

10 A Mandibular palp with regular comb of setae along distal article...................................... 11

  B Mandibular palp with irregular, non-comb-like ventral setae... 16

11 A Posterior margins of merus and carpus of P7 with long setae (longer than the spines) 12

  B Posterior margin of merus and carpus of P7 with setae, if any, shorter than the spines 14

12 A Posterodistal angle of basis of P7 without spines; lobes of telson each with at least 3 apical spines.. 13

  B Posterodistal angle of basis of P7 with 2 spines; lobes of telson each with 2 apical spines only; body banded in blue and green or black and yellow (males with curly setae) .... *Gammarus tigrinus*

13 A Inner branch of U3 <half length of outer branch; setae on merus and carpus of P7 few .... *Gammarus finmarchicus*

  B Inner branch of U3 >half length of outer branch; setae on merus and carpus of P7 dense .... *Gammarus duebeni*

14 A Inner branch of U3 <half length of outer branch; posterodistal angle of basis of P7 without spines.. *Gammarus finmarchicus*

  B Inner branch of U3 >half length of outer branch; posterodistal angle of basis of P7 with spines.. 15

15 A Anterior margin of merus and carpus of P7 with setae longer than the spines; mandibular palp with 2 or 3 groups of setae on outer surface .... *Gammarus oceanicus*

  B Anterior margin of merus and carpus of P7 without setae or with only a few short ones; mandibular palp with only 1 group of setae on outer surface (males with curly setae) .... *Gammarus chevreuxi*

16 A Mandibular palp with 1 or 2 groups of setae on outer surface; inner branch of U3 without terminal spines .... *Gammarus tigrinus*

B Mandibular palp with at least 3 groups of setae on outer surface; inner branch of U3 with terminal spines... 17

17 A Merus and carpus of P7 with abundant long setae (longer than the spines); urosome segments with long setae dorsally .... *Gammarus zaddachi*

B Merus and carpus of P7 with few short setae (± length of the spines); urosome segments with few short setae (± length of the spines).. *Gammarus salinus*

Key to the genus *Corophium*

n.b. The 4th article of the peduncle of the 2nd antenna (A2) figures prominently in the key below. In both males and females, the 2nd antenna appears to have 6 articles plus a few minute terminal ones. The two largest and longest articles are clearly the middle two, technically the 4th and 5th: the 4th is therefore the proximal of the two large middle ones, or the apparent 3rd article counting from the head.

1 A Urosome segments separate.. 2

B Urosome segments fused into single mass 5

2 A Ventrolateral angle of 4th article of A2 with 1–2 accessory spines as well as major curved spine; inner distal angle of peduncle of U2 with 2 spines .... *C. curvispinum*

B Ventrolateral angle of 4th article of A2 without accessory spines, with or without major spine; inner distal angle of peduncle of U2 without spines 3

3 A Outer margin of U1 with <9 spines, inner margin with 2–3 spines; peduncle of U3 with distolateral angle greatly produced .... *C. multisetosum*

B Outer margin of U1 with 10 or more spines, inner margin with 1 or 3–4 spines; peduncle of U3 without greatly expanded distolateral angle 4

4 A Inner margin of U1 with 1 (distal) spine; outer proximal margin of U1 with 3–4 setae and, distally, with up to 15 spines .... *C. arenarium*

B Inner margin of U1 with 3–4 spines; outer proximal margin without setae, and, distally, with up to 12 spines .... *C. volutator*

5 A Urosome with lateral flange obscuring insertions of uropods from dorsal view.................... 6

B Urosome without such lateral flange 9

6 A Ventrolateral angle of 4th article of A2 drawn out into 2 forwardly-projecting teeth of which lower is larger (i.e. specimen male) 7

B Ventrolateral angle of 4th article of A2 with spines but not drawn out into teeth (i.e. specimen female) 8

7 A Inner lower margin of 4th article of A2 with 1–3 spines proximally........................ *C. acutum*

B Inner lower margin of 4th article of A2 without spines proximally......................... *C. lacustre*

8 A Inner lower margin of 4th article of A2 with 4 large spines .... *C. acutum*

B Inner lower margin of 4th article of A2 with 1–2 small spines .... *C. lacustre*

9 A Ventrolateral angle of 4th article of A2 drawn out into 2–3 forwardly-projecting teeth of which lower/lowest is larger/largest (i.e. specimen male) 10

B Ventrolateral angle of 4th article of A2 with spines but not drawn out into teeth (i.e. specimen female) 12

10 A Outer margin of U1 peduncle with spines only distally .... *C. crassicorne*

B Outer margin of U1 peduncle without this configuration 11

11 A Inner margin of basal article of A1 with proximal process at level of tip of rostrum (view from dorsal) .... *C. insidiosum*

B Inner margin of basal article of A1 without any such protruberance .... *C. acherusicum*

12 A Spines on ventral surface of 4th article of A2 not in pairs .... *C. crassicorne*

B Spines on ventral surface of 4th article of A2 in pairs 13

13 A With at least 3 pairs of such spines .... *C. acherusicum*

B With only 1 or 2 pairs of such spines 14

14 A Inner margin of U1 peduncle with 1 spine .... *C. insidiosum*

B Inner margin of U1 peduncle with 3–4 spines .... *C. bonnellii*

Lysianassidae
Lysianassa ceratina (Fig. 75)
A common shallow-water species in the west of the region where it may be found amongst submerged vegetation in lagoons. Up to 10 mm long.

Ampeliscidae
Ampelisca brevicornis (Fig. 75)
A widespread species on shallow sand or gravel substrata that occurs in lagoons especially in the west of the region. Up to 12 mm long.

Amphilochidae
Gitana sarsi (Fig. 75)
A widely distributed species throughout the region that occasionally occurs amongst submerged vegetation in lagoons. Up to 3 mm long.

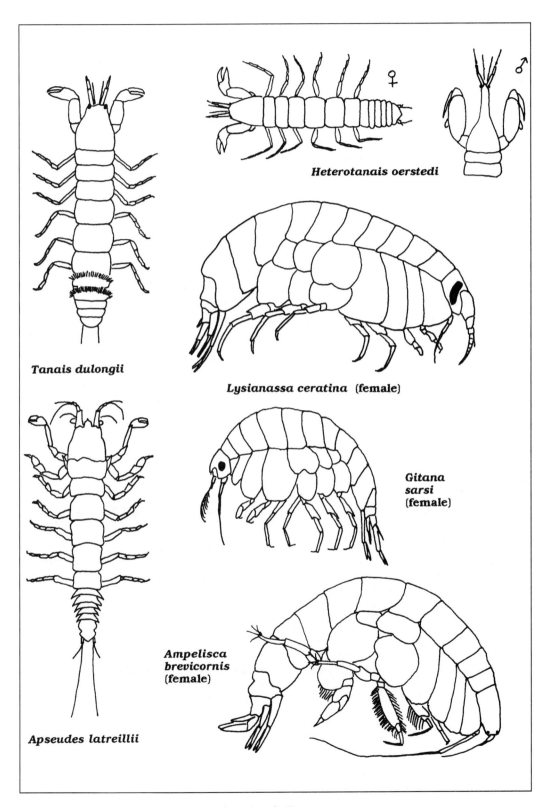

Fig. 75. Brackish-water tanaids and amphipods (I).

Leucothoidae

Leucothoe spp. (Fig. 76)

Species of *Leucothoe* are locally common components of mud and sand faunas in the region and can be expected to occur in the more saline brackish waters. Most are <8 mm long; one species achieves 18 mm length.

Stenothoidae

Metopa pusilla (Fig. 76) and *Stenothoe monoculoides* (Fig. 76)

These two stenothoids occur amongst submerged vegetation and hydroid colonies; both have occasionally been recorded from such habitats in high-salinity lagoons along the Channel. Both up to 3 mm long.

Talitridae

Orchestia spp. (Fig. 76) and *Talitrus saltator* (Fig. 76)

Talitrids are terrestrial or semiterrestrial species that only marginally belong to aquatic faunas. Nevertheless, they are common animals around the margins of estuaries; they occur in burrows or amongst drift litter at or above the high tide mark and roam onto the exposed intertidal flats during low tide. *Orchestia* species are often associated with debris on shingle, whilst *Talitrus* is more characteristic of sand. *O. remyi roffensis* is only known from the Medway Estuary, Kent. Up to 25 mm long.

Hyalidae

Hyale nilssoni (Fig. 76)

A common species thoughout the region that occurs somewhat sporadically in both estuaries and lagoons. Rocky shores, however, are its main habitat. Up to 8 mm long.

Gammaridae

Gammarus pulex (Fig. 78)

A fresh water species that occasionally gets washed or strays into dilute estuarine or lagoonal brackish waters (<6‰). Up to 30 mm long.

Gammarus locusta (Fig. 77); *Gammarus oceanicus* (Fig. 77); *Gammarus zaddachi* (Fig. 77); *Gammarus salinus* (Fig. 77)

A group of closely related species that can all occur over a salinity range of some 4 to 35‰ especially in non-tidal situations; the particularly closely related pair *G. zaddachi* and *G. salinus* extend to 1‰ in several areas. Nevertheless, as described in the text, they often subdivide a given habitat between them to avoid abortive hybridization, and the salinity gradient is often the axis of separation. In brackish waters, they occur amongst algae and other vegetation, as well as generally over the sediment surface, beneath stones, etc. All four occur in the Baltic and throughout northwestern Europe, although *G. oceanicus*, *G. salinus* and *G. zaddachi* do not occur further south than the western Channel. *G. locusta* extends southwards to Portugal. They appear to behave biologically as 'good species' but morphologically they are very similar and not always easy to distinguish. Up to 33 mm long.

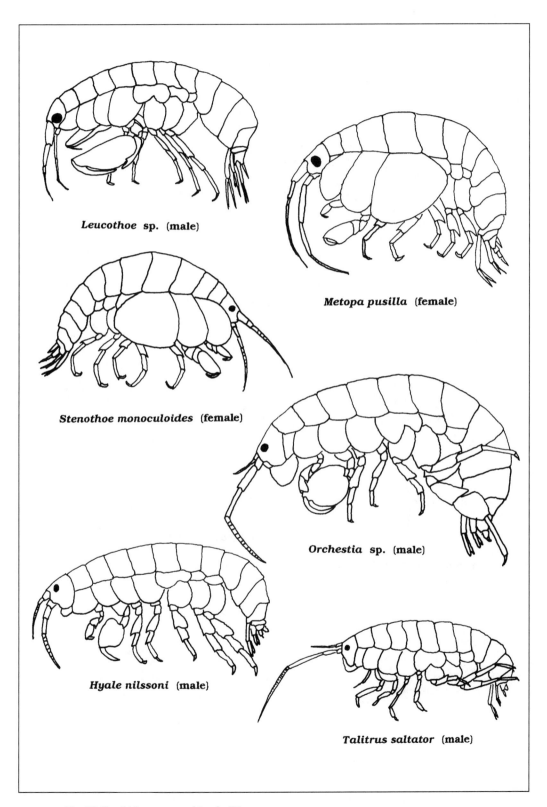

Leucothoe sp. (male)

Metopa pusilla (female)

Stenothoe monoculoides (female)

Orchestia sp. (male)

Hyale nilssoni (male)

Talitrus saltator (male)

Fig. 76. Brackish-water amphipods (II).

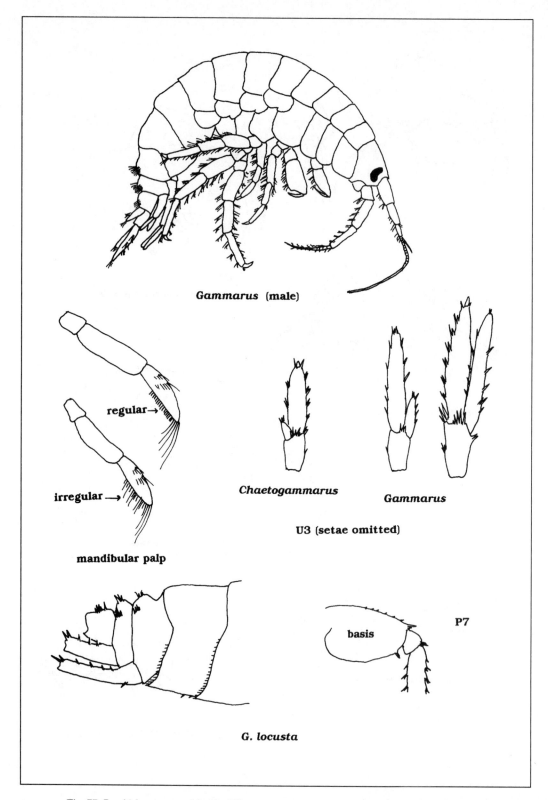

Gammarus (male)

regular→

irregular→

Chaetogammarus

Gammarus

U3 (setae omitted)

mandibular palp

basis

P7

G. locusta

Fig. 77. Brackish-water amphipods (III).

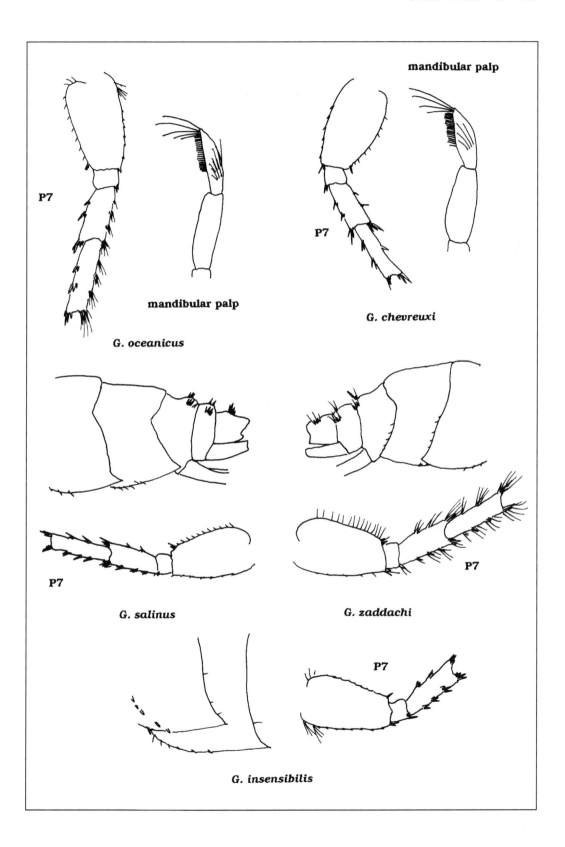

mandibular palp

P7

G. oceanicus

mandibular palp

P7

G. chevreuxi

mandibular palp

P7

G. salinus

P7

G. zaddachi

P7

G. insensibilis

Gammarus duebeni (Fig. 78)

In non-tidal habitats *G. duebeni* occurs throughout the whole salinity range of brackish water, and, especially in the west of the region but also locally elsewhere (usually in the absence of *G. pulex*), it is present in fresh water. In tidal waters, it is characteristic of the head of estuaries. High-level pools on rocky shores and fresh water discharges down marine shores are also colonized. Its distribution encompasses the Baltic, the southern and eastern North Sea and the seas around the British Isles. Up to 22 mm long.

Gammarus tigrinus (Fig. 78)

A species introduced from North America that now occurs in estuaries and, locally, in fresh waters in the west of the region and along the southern coast of the North Sea. Up to 12 mm long.

Gammarus chevreuxi (Fig. 77)

A southern species that extends up the Atlantic coast into northwest Europe as far as North Wales in the north and south Devon in the east. It has been recorded from low-salinity lagoons (some 1–10‰), brackish marshes and estuaries in southwest England and in Brittany, usually in association with the sediment (mud, sand or gravel). Otherwise it occurs southwards to Morocco. Up to 13 mm long.

Gammarus insensibilis (Fig. 77)

Essentially a Mediterranean and Black Seas species that just extends into northwestern Europe, in which it occurs in a few lagoons in southern and southeastern England (from East Anglia round to Dorset) and in Lough Hyne (Eire). Its local salinity range is 12–36‰. Up to 19 mm long.

Gammarus crinicornis (Fig. 78)

An uncommon species known from the Mediterranean to the southern North Sea that is particularly associated with clean sands but occasionally penetrates into the higher salinity reaches of estuaries as in The Netherlands. Up to 20 mm long.

Gammarus finmarchicus (Fig. 78)

A widespread North Atlantic species that extends southwards to the Western Approaches. It is the least euryhaline of the regions gammarids, but occasionally penetrates into the mouths of estuaries where it may be found in shallow pools of water with algae. Up to 20 mm long.

Chaetogammarus marinus (Fig. 78)

Formerly known as *Marinogammarus,* this is a widespead northeast Atlantic species that is typically associated with fucoid algae growing on muddy shingle, and it occurs in such habitats in a few lagoons and in the mouths of estuaries, as well as in sheltered marine habitats generally. It occurs throughout the region. Up to 25 mm long.

Chaetogammarus stoerensis (Fig. 78) and *Chaetogammarus pirloti* (Fig. 78)
Formerly known as *Marinogammarus,* these two species occur sporadically in areas of fresh water discharge, especially over intertidal shingle or coarse gravel. *C. pirloti* (up to 14 mm long) is restricted to the north and west, whereas *C. stoerensis,* although being mainly western, is also known from sporadic localities throughout the region. It is only up to 8 mm in length.

Eulimnogammarus obtusatus (Fig. 78)
Formerly known as *Marinogammarus,* this species is commonly associated with algae on relatively clean gravel and shingle throughout the region, and can occur in the mouths of estuaries and estuarine lagoons, as well as in the barrier system of high-salinity lagoons. Up to 20 mm long.

Pectenogammarus planicrurus (Fig. 78)
A specialist inhabitant of clean shingle from the Mediterranean northwards to the Isle of Man and the southern North Sea, this species may occur in the barrier system of lagoons and in the spits at the mouths of estuaries. Up to 9 mm long.

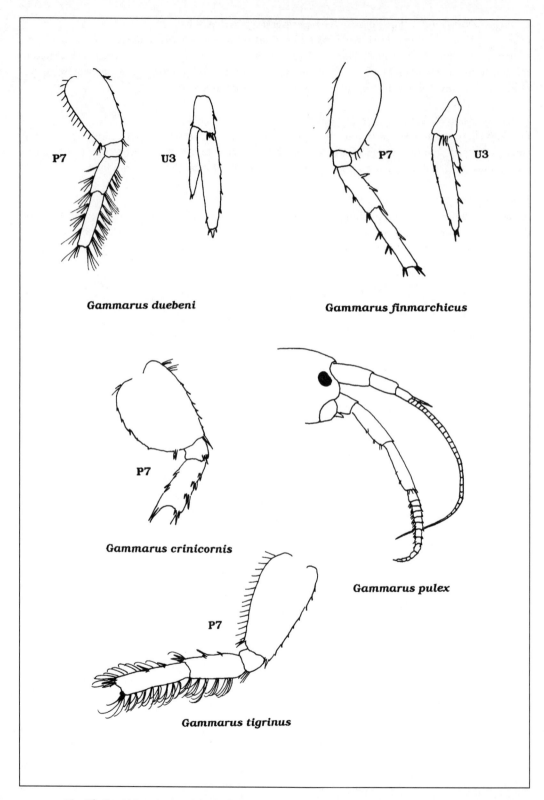

Fig. 78. Brackish-water amphipods (IV).

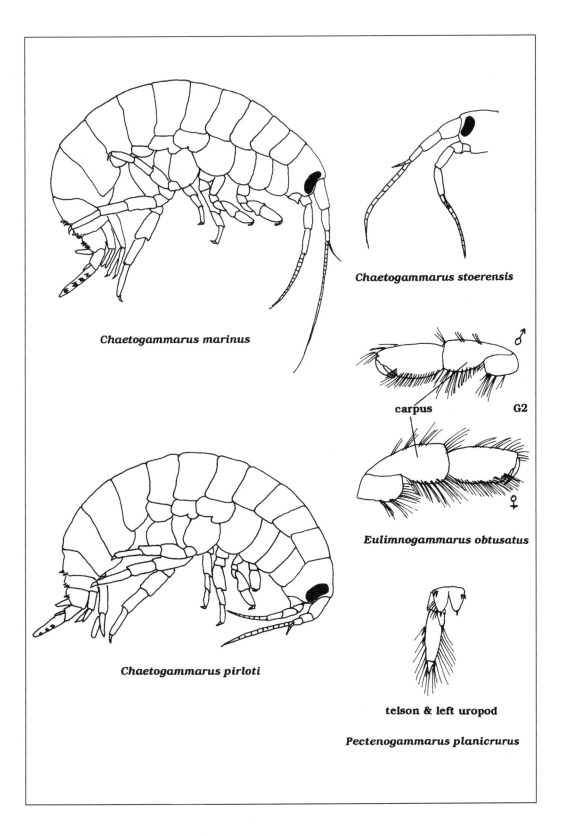

Chaetogammarus marinus

Chaetogammarus stoerensis

carpus ♂ G2

Eulimnogammarus obtusatus ♀

Chaetogammarus pirloti

telson & left uropod

Pectenogammarus planicrurus

Melitidae

Melita palmata (Fig. 79)

A common member of brackish-water (and marine) faunas occurring down to salinities of some 5–6‰ in lagoons and estuaries (and in the Baltic). It is usually associated with soft sediments that include surface stones or shingle, and can be found throughout the region. Up to 16 mm long.

Melita pellucida (Fig. 79)

M. pellucida is more restricted to brackish waters than its congener above, but is less widespread and less common. It has been recorded sporadically from Norway to the French Channel coast, usually in ditches or harbours. Up to 6 mm long.

Cheirocratus sundevalli (Fig. 79)

A widespread species occurring throughout the region and occasionally recorded from muddy sand in lagoons. Up to 10 mm long.

Gammarella fucicola (Fig. 79)

Almost always associated with algae, this species has occasionally been recorded from the mouths of estuarine lagoons in the west of the region. Up to 10 mm long.

Maera grossimana (Fig. 79)

A widely distributed species that extends into the southwest of the region (from the Bristol Channel to the Isle of Wight). It may occur beneath stones in the mouths of estuaries. Up to 10 mm long.

Haustoriidae

Bathyporeia pilosa (Fig. 79)

Although restricted to fine and medium sandy sediments, this species can occur down to some 4‰ salinity wherever suitable substrata are present. It occurs in the Baltic and at high tidal levels in sandy estuaries throughout the region except in the extreme north. Up to 6 mm long.

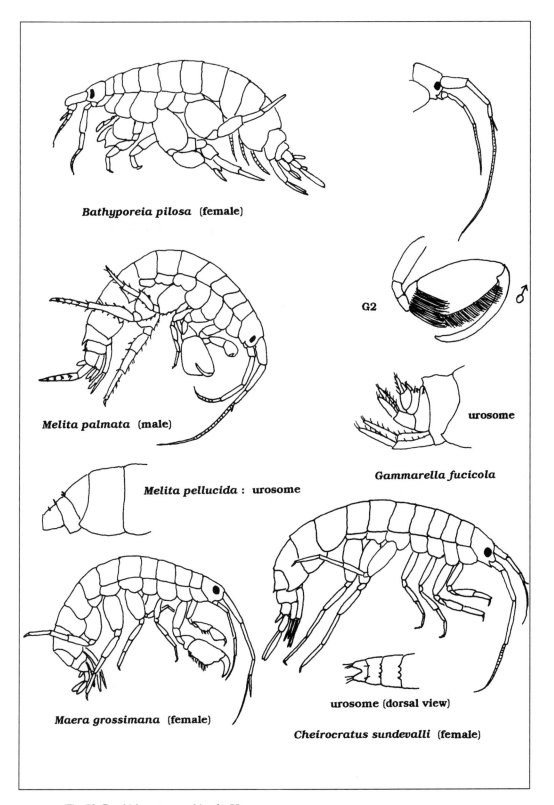

Fig. 79. Brackish-water amphipods (V).

Pontoporeia femorata (Fig. 80)
A member of an Arctic and Baltic genus that just extends into the region in the extreme east (Denmark), where it occurs in salinities of down to some 6‰. Up to 10 mm long.

Haustorius arenarius (Fig. 80)
Like *Bathyporeia pilosa* above, an inhabitant of sands that can occur in low salinities (<10‰) wherever medium to coarse sands are present in estuaries (for example, in the Severn). It occurs throughout the region, but since its preferred sediment is uncommon in estuaries it has rarely been recorded from brackish waters (except in the Baltic). Up to 13 mm long.

Urothoe spp. (Fig. 80)
Various *Urothoe* species penetrate into the mouths of estuaries where they burrow into the surface layers of soft sediments; however, they are not regular members of estuarine faunas. Up to 8 mm long.

Oedicerotidae
Pontocrates spp. (Fig. 80)
Pontocrates species are associated with sandy sediments throughout the region, and they are likely to occur in the more saline of brackish waters. Up to 7 mm long.

Phoxocephalidae
Phoxocephalus holbolli (Fig. 80)
A northern species extending as far south as the Channel. It occurs on fine sediments and since it is present in the Baltic it is likely that it will be found in northwest European estuarine or lagoonal brackish waters. Up to 7 mm long.

Liljeborgiidae
Liljeborgia kinahani (Fig. 80)
An uncommon species associated with coarse sediments that occurs sporadically throughout the region, and has once been recorded from a lagoon in the central region of the Channel. Up to 3 mm long.

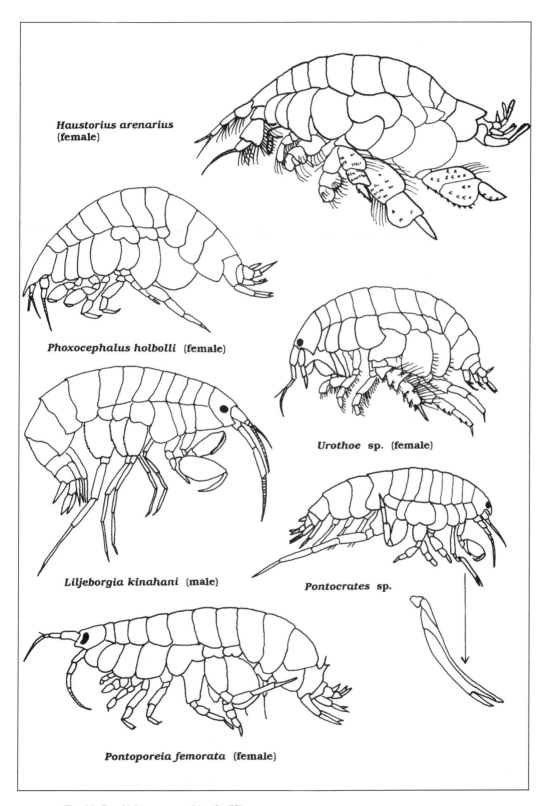

Fig. 80. Brackish-water amphipods (VI).

Calliopiidae

Apherusa spp. (Fig. 81)

Apherusa contains several Arctic and North Atlantic species that occur amongst algae, and they are likely to occur in the more saline of brackish waters. Up to 8 mm long.

Calliopius laeviusculus (Fig. 81)

A widespread species that in the extreme west of the region can occur in salinities of down to 6‰ but that rarely does so elsewhere. Occasionally recorded from lagoons along the Channel coast. Up to 14 mm long.

Gammarellus angulosus (Fig. 81)

An Arctic and North Atlantic species that occurs uncommonly in lagoon and lagoon-like habitats (e.g. Lough Hyne) and the mouths of estuaries. Up to 15 mm long.

Atylidae

Atylus swammerdami (Fig. 81)

A ubiquitous and locally common species amongst vegetation, especially over a sandy substratum. It has occasionally been recorded from lagoons. Up to 10 mm long.

Dexaminidae

Dexamine spinosa (Fig. 81)

An Atlantic species occurring from Norway to West Africa and sporadically recorded from amongst *Zostera* or *Ruppia* in lagoons and from amongst algae in the mouths of estuaries. Up to 14 mm long.

Ampithoidae

Ampithoe rubricata (Fig. 81)

A tube-building species that occurs throughout the region and may occasionally be found building its mud tubes on stones or submerged vegetation in the more marine of brackish waters. Up to 20 mm long.

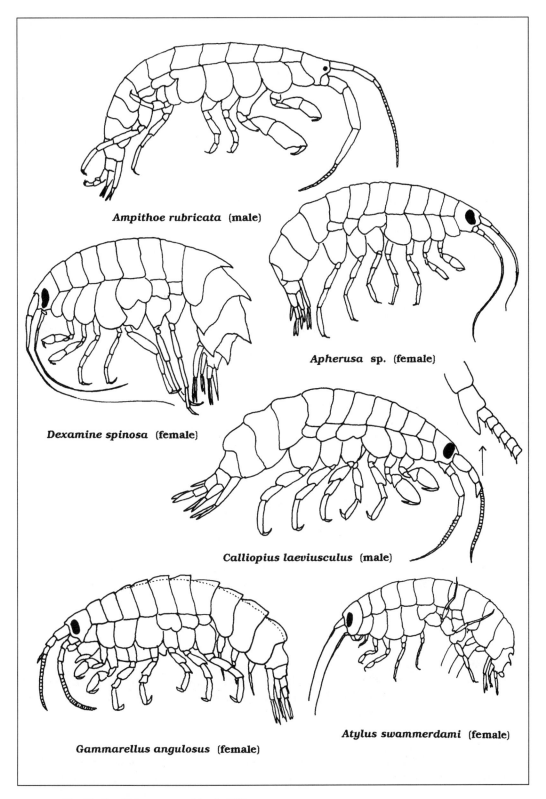

Ampithoe rubricata (male)

Apherusa sp. (female)

Dexamine spinosa (female)

Calliopius laeviusculus (male)

Gammarellus angulosus (female)

Atylus swammerdami (female)

Fig. 81. Brackish-water amphipods (VII).

Aoridae

Aora typica (Fig. 82)

Also recorded as *A. gracilis*, this cosmopolitan species builds mud tubes amongst hydroids and submerged vegetation, and is able to withstand slight dilution of its medium and occur in lagoons, for example along the Channel coast of England. Up to 9 mm long.

Lembos websteri (Fig. 82)

A North Atlantic species associated with algae and other submerged vegetation that has occasionally been recorded from lagoons. Up to 6 mm long.

Leptocheirus pilosus (Fig. 82)

A common and widespread brackish-water species, particularly (but not exclusively) in non-tidal habitats, that constructs mud tubes on the sediment surface, on submerged vegetation (e.g. *Ruppia*), amongst hydroid colonies (e.g. *Cordylophora*), etc., in salinities of down to 5‰ and locally less. It occurs from the Baltic westwards throughout the region. Up to 5 mm long.

Microdeutopus gryllotalpa (Fig. 82)

A common inhabitant of salt-marshes, lagoons and equivalent habitats throughout the region (and on southwards to the Mediterranean and Black Seas) most commonly in salinities of above 18‰ but on occasion down to some 10‰. In lagoons, it constructs mud tubes on stones and shells beneath a cover of *Zostera, Ruppia* or other vegetation and sometimes on the vegetation itself. Densities of 13 000 individuals/m^2 have been recorded. It also occurs in rock pools and docks. Up to 10 mm long.

Isaeidae

Gammaropsis maculata (Fig. 82)

A North Atlantic species present throughout the region in association with hydroids and algae. It has occasionally been recorded from lagoons. Up to 10 mm long.

Microprotopus maculatus (Fig. 82)

A small species distributed from Norway to the Mediterranean and occurring on sandy sediments. It has been recorded from a southern North Sea lagoon. Up to 3 mm long.

Corophiidae

Corophium volutator (Fig. 83)

A widespread and often abundant burrower in muddy sediments in estuaries, lagoons, ditches, salt-pans and all other forms of brackish habitat, as well as in sheltered marine areas, throughout the region and beyond (e.g. in the Mediterranean and in Atlantic North America). It can occur down to 1‰ salinity, in both lagoons and estuaries. Densities of this deposit- and suspension-feeding amphipod can approach 100 000 individuals/m^2. Up to 10 mm long.

Corophium arenarium (Fig. 83)

A similar species to *C. volutator* above, that replaces it in sandy estuarine and lagoonal sediments over a salinity range of some 6-35‰. Its apparently patchy distribution (southern North Sea, northern Ireland and Irish Sea) may be an artifact of past confusion with the species above. Up to 7 mm long.

Corophium insidiosum (Fig. 83)

A widespread but local inhabitant of non-tidal brackish waters in which it constructs tubes of mud on hydroid colonies or submerged vegetation. It occurs from Denmark through the southern North

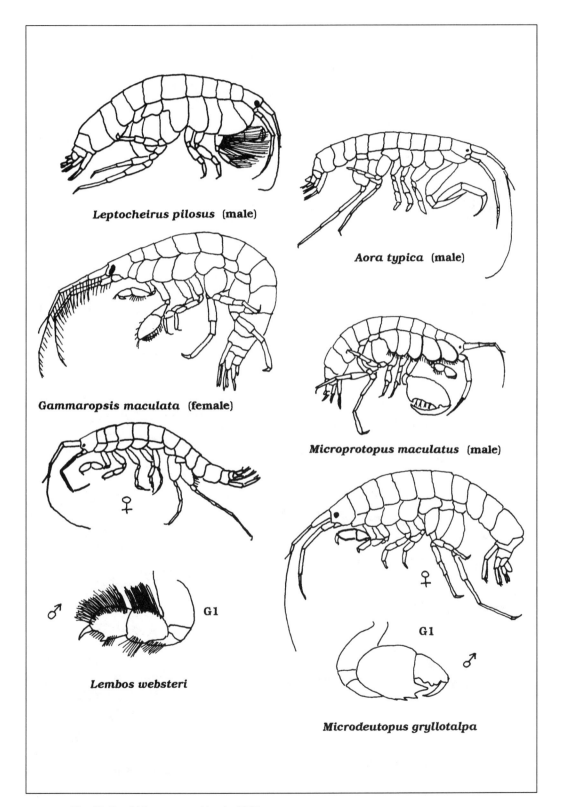

Leptocheirus pilosus (male)

Aora typica (male)

Gammaropsis maculata (female)

Microprotopus maculatus (male)

♀

Lembos websteri

♂ G1

♀

G1 ♂

Microdeutopus gryllotalpa

Fig. 82. Brackish-water amphipods (VIII).

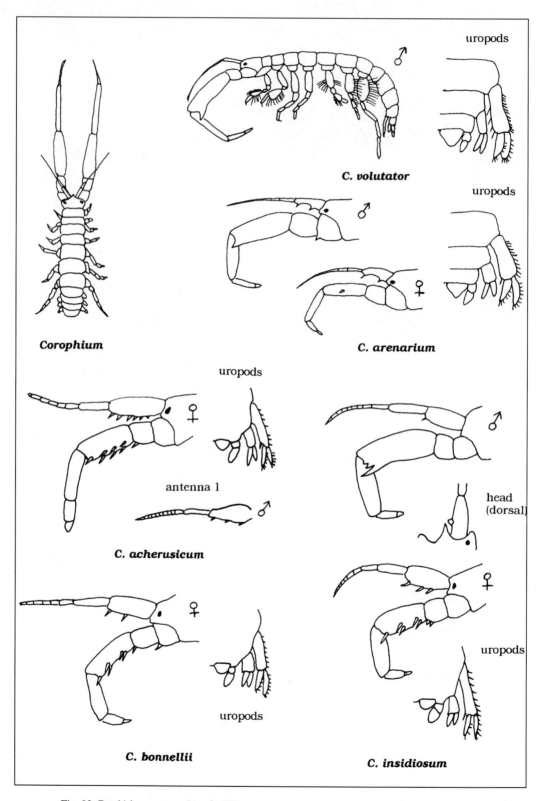

Fig. 83. Brackish-water amphipods (IX).

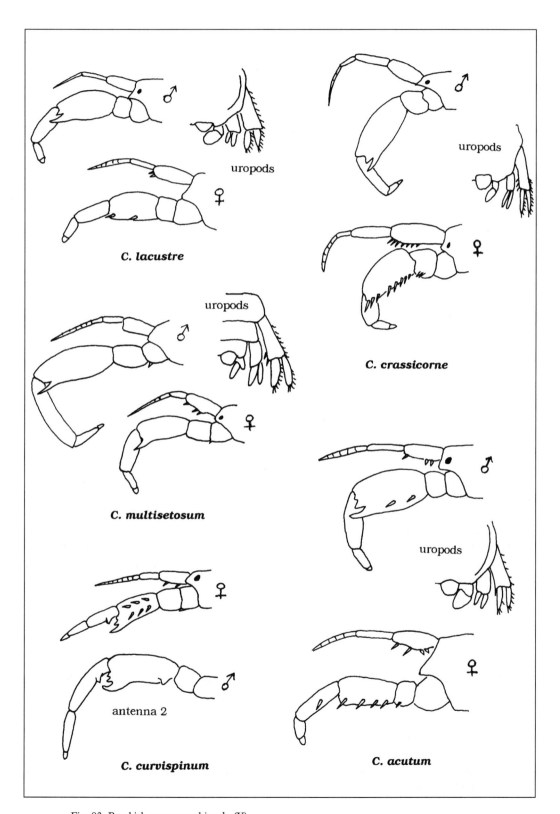

Fig. 83. Brackish-water amphipods (X).

Sea and Channel to Brittany and on to the Mediterranean, in waters (usually) of 12–35‰. Densities of 12 000 individuals/m² have been recorded. Up to 5 mm long.

Corophium lacustre (Fig. 83) and *Corophium multisetosum* (Fig. 83)
These species, like most *Corophium,* construct mud tubes on hydroids, submerged vegetation and other surfaces, although *C. multisetosum* can also burrow in soft sediments. They occur in fresh or slightly brackish waters, up to recorded maxima in the vicinity of 16‰ salinity, in the Baltic, the southern North Sea, the Channel, and southern Welsh coasts. (There is one record of *C. lacustre* from a Hampshire lagoon in full-strength sea water, although this may be based on a misidentification.) *C. lacustre* is up to 6 mm long; *C. multisetosum* is half as long again.

Corophium curvispinum (Fig. 83)
A widespread European species found in river systems discharging into the Caspian, Black, Baltic and North Seas, and occurring in fresh and brackish waters. In northwest European brackish waters it is restricted to the mouths of German rivers. Up to 6 mm long.

Corophium acherusicum (Fig. 83)
A cosmopolitan builder of mud tubes on submerged vegetation, hydroids, hard surfaces, etc. that occurs in estuaries and harbours in the southern half of the region. It can tolerate 20‰ and probably lower salinities. Up to 5 mm long.

Corophium crassicorne (Fig. 83)
A widespread species that burrows in soft sediments and tends to replace *C. volutator* in the subtidal zone of estuaries and marine waters over the western half of the region. Up to 5 mm long.

Corophium bonnellii (Fig. 83)
The least euryhaline of the brackish-water *Corophium* species, *C. bonnellii* can penetrate the mouths of estuaries and build its mud tubes on hydroids, stones, algae, etc. in subtidal regions. In northwest Europe, it is known mainly from the western half of the region, but it is also present in the southern North Sea. Up to 6 mm long.

Corophium acutum (Fig. 83)
A cosmopolitan species that in northwest Europe builds its tubes on artificial substrata (buoys, pilings, etc.) in docks and harbours exposed to rapid water movements, including in the mouths of estuaries. Occurs in the south-west of the region (from Southampton Water to southwest Eire). Up to 4 mm long.

Cheluridae
Chelura terebrans (Fig. 84)
A widespread species that bores into waterlogged timber, including into wooden piers, posts, etc. in the mouths of estuaries. There is one record from a Channel coast lagoon. Up to 6 mm long.

Ischyroceridae
Erichthonius spp. (Fig. 84)
The distributions of the individual species of *Erichthonius* are poorly known as a result of past lumping together under the one name, *E. brasiliensis.* They build mud tubes amongst hydroids and on algae and *Zostera,* and have been recorded from lagoons along the Channel coast, most notably The Fleet, Dorset. Up to 10 mm long.

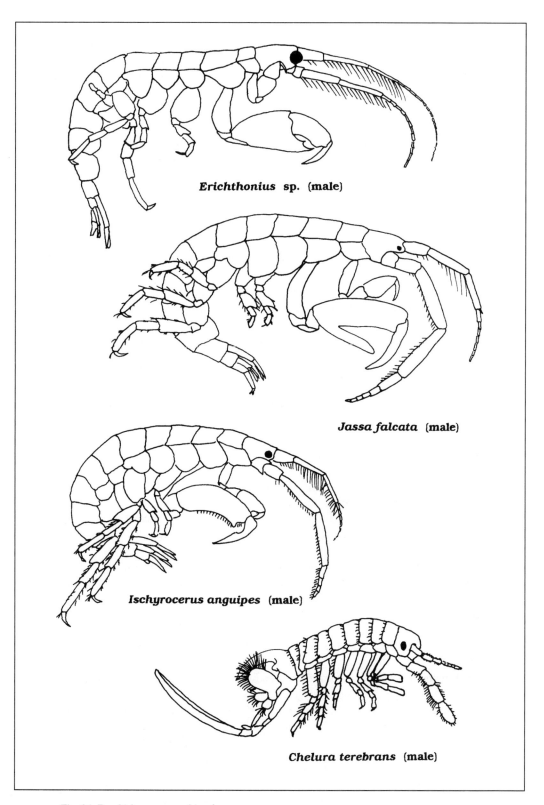

***Erichthonius* sp. (male)**

***Jassa falcata* (male)**

***Ischyrocerus anguipes* (male)**

***Chelura terebrans* (male)**

Fig. 84. Brackish-water amphipods.

Jassa falcata (Fig. 84)

A cosmopolitan species that constructs mud tubes amongst hydroids and algae or on man-made structures such as buoys, pilings, ships and the cooling-water systems of power stations. Occurs throughout the region, including in harbours in the mouths of estuaries and in lagoons of the Channel coast (in salinities down to some 18‰). Up to 12 mm long.

Ischyrocerus anguipes (Fig. 84)

Like *Jassa falcata* above, *Ischyrocerus* builds mud tubes on man-made buoys, pilings, etc. and amongst hydroids and algae, and is similarly responsible for marine fouling. It occurs throughout the region, including in ports and harbours in the mouths of estuaries. Up to 15 mm long.

Decapods (crabs and prawns)

Crabs need no introduction, but prawns may be distinguished from the somewhat similar mysids by their larger size (up to 110 mm long) and by the stouter (effectively uniramous) pereiopods and carapace (which covers the whole of the thorax and lacks the mysid groove). Smaldon (1979) and Ingle (1983) describe British species. See Fig. 85.

1 A Crab-shaped (i.e. small abdomen reflexed back under large cephalothoracic carapace) 2

  B Prawn- or shrimp-shaped (i.e. large abdomen, ending in a tail fan, not hidden by the carapace).. 4

2 A Outer face of chela with dense felt of setae; area of carapace between eyes ('frontal region') drawn out into four sharp points .... *Eriocheir sinensis*

  B Outer face of chela without mat of hairs; frontal region of carapace not drawn out into four points 3

3 A Anterolateral margins of carapace with 5 teeth; frontal region of carapace 5-pointed, the median one projecting furthest; carapace up to 60 mm long .... *Carcinus maenas*

  B Anterolateral margins of carapace with 4 teeth; frontal region of carapace with 2, flat lobes; carapace up to 12 mm long .... *Rhithropanopeus harrisii*

4 A First leg ending in a sub-chela (pincer in which the movable claw articulates with the fixed one distally); body dorso-ventrally flattened.. *Crangon crangon*

  B First leg chelate (bearing a pincer in which the movable claw articulates with the fixed one proximally, like that of a crab); body not dorso-ventrally flattened 5

5 A Chela very long and slender, with comb-like cutting edges; rostrum very short not projecting beyond eyes .... *Pasiphaea sivado*

  B Chela without comb-like cutting edges; rostrum large, projecting well in front of eyes 6

6 A Only one of the teeth on the dorsal margin of the rostrum situated behind the level of the posterior edge of the orbit (although a second may be immediately above that level) 7

  B 2 or 3 dorsal teeth on the rostrum situated behind the level of the posterior edge of the orbit 8

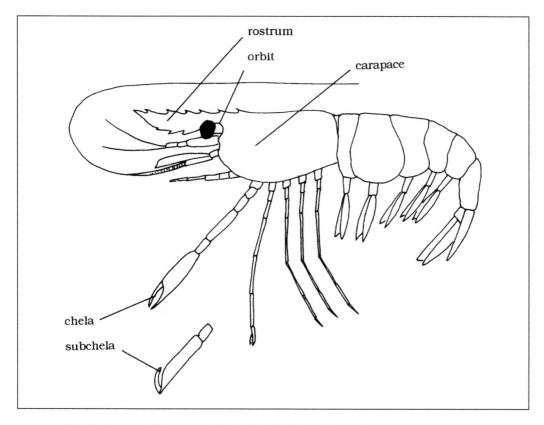

rostrum

orbit

carapace

chela

subchela

Fig. 85. Anatomical features used in the identification of decapod crustaceans.

7 A Yellowish grey in colour, with red spots on the lower half of the rostrum; one dorsal rostrum tooth clearly behind posterior margin of orbit, a second immediately above that margin .... *Palaemon adspersus*
  B Transparent, without red spots on the rostrum; with 1 dorsal rostrum tooth clearly behind posterior margin of orbit and without one immediately above that margin .... *Palaemonetes varians*

8 A 3 dorsal rostral teeth behind level of posterior margin of orbit....................... *Palaemon elegans*
  B 2 dorsal rostral teeth behind level of posterior margin of orbit....................................... 9

9 A Rostrum straight, with dorsal teeth present on its distal third .... *Palaemon longirostris*
  B Distal third of rostrum upturned, without dorsal teeth............................... *Palaemon serratus*

Pasiphaeidae
Pasiphaea sivado (Fig. 86)
A western species absent from the Channel and North Sea that may penetrate into the brackish-water stretches of some of the larger estuaries of the region's west coasts. Up to 100 mm long.

Palaemonidae
Palaemonetes varians (Fig. 86)
This most characteristic brackish-water species occurs in salinities of down to 1‰ (and temporarily in fresh water) in lagoons, equivalent ponds and ditches, and more rarely in salt-marsh pools, throughout the region except in the extreme north; it has very rarely been recorded from the sea although it does occur in lagoonal habitats permanently experiencing salinities of over 30‰. It is almost ubiquitous amongst *Phragmites, Ruppia, Zostera* and in open habitats, and it also inhabits completely land-locked systems. Its diet appears mainly to be benthic diatoms and detritus, although some populations at least consume small animals, including the rare anemone *Nematostella*. Up to 50 mm long.

Palaemon longirostris (Fig. 86)
A prawn which reaches its northern limit in southern England and the southern shores of the North Sea, and only sporadically known from the region, mainly from Dutch and East Anglian estuaries and associated lagoon-like habitats (but also from estuaries along the south and southwestern coasts of Britain). As adults they may extend into fresh waters but must migrate into the estuaries to breed. The larvae are marine and baby prawns eventually pass back into the estuaries (and on into fresh water) when some 6 mm long. Up to 80 mm long.

Palaemon elegans (Fig. 86)
A widespread species in Europe, known in the older literature as *P.* (or *Leander*) *squilla*, that occurs in salinities of below 4‰ in some lagoons of the Black Sea basin and of below 6‰ in a Danish lagoon, but which is nevertheless mainly marine in its northwest European distribution. It is characteristic of areas of open habitat or with macroalgae. Up to 65 mm long.

Palaemon adspersus (Fig. 86)
In the Baltic, this species penetrates to below 5‰, and forms the basis of an important fishery in the Danish Belt Sea; otherwise, however, it is uncommon in northwestern Europe. Like several other prawns it is known to winter in offshore, deeper and more saline regions and to enter coastal brackish waters only during the summer months. In Denmark it is especially associated with *Zostera* beds, and is replaced by *P. elegans* in areas or at times of no eelgrass cover. Up to 70 mm long.

Palaemon serratus (common prawn) (Fig. 86)
Penetrates into the mouths of some estuaries, but it is the least brackish-water of the region's palaemonids. Up to 110 mm long.

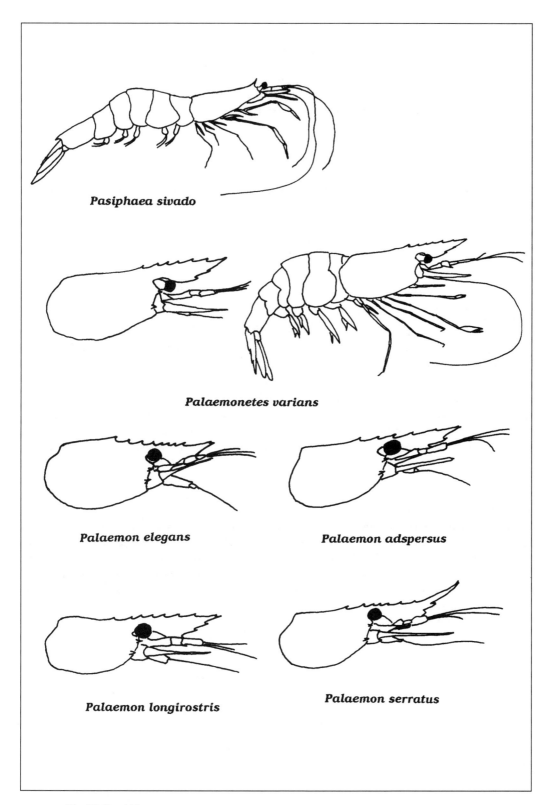

Pasiphaea sivado

Palaemonetes varians

Palaemon elegans

Palaemon adspersus

Palaemon longirostris

Palaemon serratus

Fig. 86. Brackish-water prawns.

Crangonidae

Crangon crangon (common or brown shrimp) (Fig. 87)

Occurs down to some 6‰ in the Baltic, but associated with higher salinities in the estuaries of the rest of northwest Europe. It is extensively fished along the southern and eastern shores of the North Sea and in the Irish Sea. Typically an animal of unvegetated sandy sediments, in which up to 55 shrimps can be present per square metre of substratum surface, but it also occurs where the ground is muddy. Up to 90 mm long.

Portunidae

Carcinus maenas (shore crab) (Fig. 87)

A common animal in estuaries and in those largely land-locked brackish ponds and lagoons that communicate with the sea. The adults can withstand salinities as low as 4‰, but the females must migrate to sea to carry the eggs since the latter are killed by concentrations of less than 20‰. Migration into deeper, saltier waters takes place during winter, especially in the north and east of the region, and a degree of migration up and down estuaries with the state of the tide also occurs. The crab, termed *Carcinides* in much of the continental literature, is omnivorous, picking over the sediment surface, for example, and opening up mussels where these are available.

Grapsidae

Eriocheir sinensis (Fig. 87)

An immigrant Chinese crab that was first noticed in the Weser in 1912 and has now spread southwestwards to the Atlantic coast of France and northeastwards into the Baltic; to date only some 20 individuals have been reported from the British Isles (all in the Thames and Humber). Adults are fresh water in habitat, although they must migrate to the sea to mate and to release the larvae; they generally congregate in the mouths of estuaries from autumn to spring whilst doing so. Young crabs therefore migrate up the estuaries and adult females migrate down them, both movements taking place in or near the month of August. Adult females usually die after carrying the eggs. The carapace is up to 60 mm long.

Xanthidae

Rhithropanopeus harrisi (Fig. 87)

A small immigrant American brackish-water species, which when it was found in the nineteenth century in the former Zuiderzee was described as new under the name *Pilumnus tridentatus*. It now occurs sporadically from Normandy eastwards along the southern North Sea coast into the Baltic (as well as in the Mediterranean and Black Seas) in salinities of down to 1‰, and the crab can temporarily inhabit, though not breed in, fresh water. It has been recorded under a variety of generic names: *Heteropanope, Pilumnopeus* and now *Rhithropanopeus.*

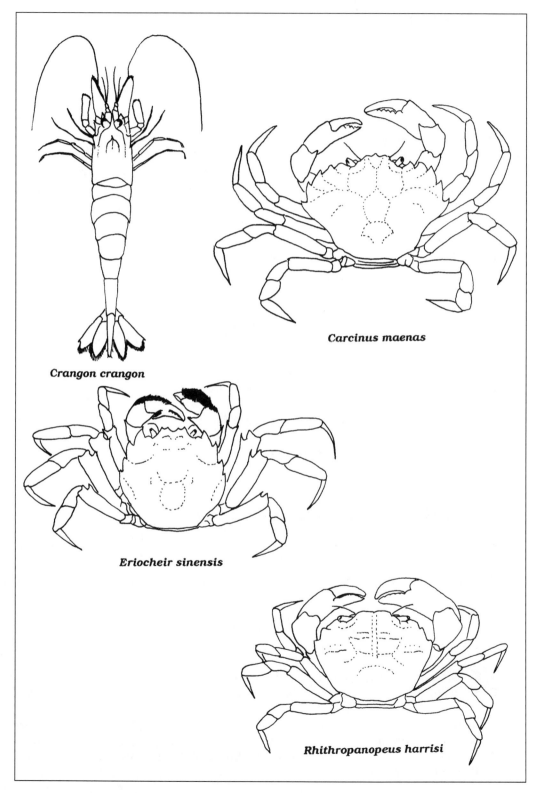

Crangon crangon

Carcinus maenas

Eriocheir sinensis

Rhithropanopeus harrisi

Fig. 87. Brackish-water shrimps and crabs.

Insects

Larvae, nymphs and wingless adults

A wide variety of insect larvae occur in brackish waters, especially in the sediment or associated with submerged vegetation. All are variants on the familiar segmented maggot, grub and caterpillar themes, and possess between 0 and 3 pairs of jointed legs as well as, in some, various forms of abdominal appendages serving a respiratory function. Most, however, cannot be identified to species; neither can the majority of insect nymphs. The key below will nevertheless proceed as far as possible, and it will include the adults of the few brackish-water wingless insects, which may appear to be larvae. Further information on fly larvae is given by Smith (1989) and on beetle larvae by Richoux (1982) and Klausnitzer (1991). Although it is not devoted to Europe, Merritt & Cummins (1984) is nevertheless a useful key to aquatic insect larvae. See Fig. 88.

1 A Without pair of jointed legs on each thoracic segment (fly larvae) 2
  B With pair of jointed legs on each thoracic segment (n.b. the legs are very small in a grub (a beetle larva) that lives attached to submerged vegetation by a pair of terminal spines).......... 10

2 A Body comprising head, 3 thoracic segments and abdomen of at least 9 segments 3
  B Body comprising head, 3 thoracic segments and 8 abdominal segments............................ 6

3 A Head can be retracted into thorax; last segment with a pair of spiracles surrounded by fleshy lobes .... tipulids
  B Head not retractile; last segment not as above 4

4 A Abdomen elongate; posterior margins of some segments with a ring-like thickening; with a long, thin, retractile respiratory tube .... ptychopterids
  B With or without respiratory tube, but if present, it is short and not rectractile 5

5 A With a pair of soft, hydraulic prolegs on 1st thoracic and last abdominal segment; often red.. chironomids
  B Without prolegs; never red.. psychodids

6 A With a distinct terminal respiratory tube that does not end in a circlet of large setae 7
  B Without a respiratory tube or, if one is present, then it ends in a circlet of plumose or pinnate setae 8

7 A Respiratory tube elongate, rectractile and unforked... syrphids
  B Respiratory tube forked.. ephydrids

8 A Head sharply demarkated from body: posterior border of head < half anterior border of 1st thoracic segment; with or without respiratory tube capped by ring of setae...... stratiomyiids
  B Head not sharply demarkated from body, so that anterior end of body tapers to maggot-like point; no respiratory tube 9

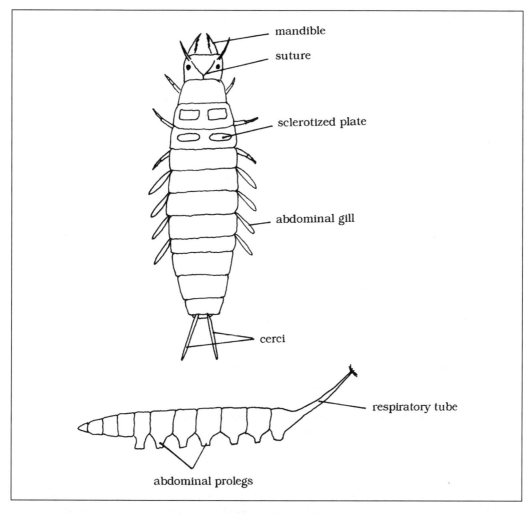

Fig. 88. Anatomical features used in the identification of insect larvae.

9 A With 4+ short, proleg-like papillae on most abdominal segments; posterior end not drawn
out into 4 processes.. tabanids

B Without proleg-like papillae, although ventral ridges may be present; posterior end drawn
out into 4 processes of which ventral ones largest and furthest projecting.......... dolichopodids

10 A Animals blue to blue-black; abdomen with 6 segments; occurring on intertidal mud
.... *Anurida maritima*

B Animals not blue; abdomen with >6 segments 11

11 A Animals living in a case of sand grains and/or vegetable debris which they carry around
with them (caddisfly larvae)... 12

B Animals not living within a movable case (although 2 spp. secrete a silken tube that may
incorporate pieces of waterweed: these can be distinguished from caddisfly larvae by their
possession of 4 pairs of abdominal prolegs)... 14

12   **A**   Dorsal surface of 3rd thoracic segment with several small sclerotized plates; 2nd and 3rd legs subequal in length .... *Limnephilus affinis*

      B   Dorsal surface of 3rd thoracic segment without sclerotized plates; 3rd leg much longer than 2nd 13

13   **A**   Head with dark longitudinal black bands .... *Mystacides longicornis*

      B   Head with dark spots .... *Oecetis ochracea*

14   **A**   With 3 long, subequal, filamentous or leaf-like, terminal processes that are > 0.3× body length 15

      B   Without 3 such projections, although other numbers and/or other types of processes may be present 17

15   **A**   Animal living in air; terminal projections filamentous; without abdominal gills .. *Petrobius brevistylis*

      B   Animal aquatic 16

16   **A**   Abdomen with 7 pairs of dorso-lateral gills, 1st pair single, the others double ..... *Cloeon dipterum*

      B   Abdomen without gills .... *Ischnura elegans*

17   **A**   Legs long, hind pair reaching beyond tip of abdomen 18

      B   Hind legs not reaching beyond tip of abdomen 22

18   **A**   Head longer than wide; animal living on the water surface........................... *Gerris thoracicus*

      B   Head wider than long; animal aquatic 19

19   **A**   Abdomen ending in 5 short spines; hind legs without fringes of setae; animal crawls 20

      B   Abdomen without 5 terminal spines; hind legs fringed by setae; animal swims using legs as oars 21

20   **A**   Abdominal segments 5–8 each with backwardly projecting mid-dorsal spine, segment 9 with large lateral spines; sides of head converging posteriorly .... *Sympetrum sanguineum*

      B   Abdominal segments without mid-dorsal or lateral spines; sides of head ± parallel .... *Orthetrum cancellatum*

21   **A**   Animal swims on its back; head with pointed ventral beak; tarsus of hind leg not enlarged .... notonectid nymph

      B   Animal swims dorsal surface uppermost; head without pointed beak; tarsus of hind legs stouter than other articles .... corixid nymph

22   **A**   Abdomen with 4 pairs of prolegs (on segments 3–6); animals green to brown, with darker mid-dorsal line and brown head; inhabit silken tube amongst waterweeds (pyralid moth caterpillars) 23

      B   Animals not as above; abdomen without prolegs or with 5 pairs (on segments 3–7) 24

23   **A**   Animal pale green .... *Acentropus niveus*

      B   Animal olive-brown .... *Nymphula nympheata*

24 A Animals in or on intertidal estuarine sediments 25
   B Animals permanently submerged in pond-like habitats 28

25 A Surface dwelling; body short, oval; legs long, fore pair extending further forwards than head, hind pair extending further backwards than end of abdomen .... *Saldula palustris*
   B Within the sediment; body elongate; legs short, not spanning a greater length than the body
   26

26 A Legs with 5 articles and with paired terminal claws.. carabids
   B Legs with 3–4 articles and with single terminal claws... 27

27 A 9th abdominal segment with 2 cerci .... *Bledius* spp.
   B 9th abdominal segment without cerci .... *Heterocerus* spp.

28 A Fat, curved, grub-like larvae with very short legs attached to submerged vegetation by pair of dorsal spines on last segment.. *Macroplea mutica*
   [Often recorded under the generic name Haemonia.]
   B Free-living animals not attached to vegetation by spines 29

29 A Legs with 5 articles and with 1–2 terminal claws... 30
   B Legs with 4 articles and with 1 terminal claw 40

30 A Abdomen with >8 segments 31
   B Abdomen with 8 segments 32

31 A Abdomen with 1 pair of feather-like lateral gills on all but last 2 abdominal segments, with 2 pairs of gills on penultimate segment, and 4 small hooks on the last; legs with 2 claws
   .... *Gyrinus* spp.
   B Abdomen without lateral gills, ending in long tapering 'tail'; legs with 1 claw .... *Haliplus* spp.

32 A Body cylindrical, ending in 3 short points (pointed last segment used for attachment to submerged vegetation) and 2 short cerci; legs short, stout, used in burrowing... *Noterus clavicornis*
   B Body tapering posteriorly, ending in 2 cerci; legs not short, stout and adapted for burrowing... 33

33 A Head not triangular, i.e. without median frontal projection 34
   B Head triangular, pointed anteriorly 37

34 A Last 2 abdominal segments with fringe of setae .... *Dytiscus circumflexus*
   B Last 2 abdominal segments without fringe of setae.. 35

35 A The 2 terminal cerci with setae at their tip and in a tuft near their base .... *Agabus* spp.
   B The 2 terminal cerci with fringe of setae along their whole length 36

36 A Dorsal surface of 1st thoracic segment rectangular .... *Colymbetes fuscus*
   B Dorsal surface of 1st thoracic segment ± triangular, lateral margins converging anteriorly
   .... *Rhantus exsoletus*

37 A Last abdominal segment a short equilateral triangle, the 2 long terminal cerci (approx. 6× length of last segment) with fringe of short setae along whole length .... *Potamonectes depressus*

   B Last abdominal segment distinctly longer than wide, the short terminal cerci (subequal to length of last segment) with scattered long setae 38

38 A Terminal projection of last segment (i.e. region posterior to insertion of the 2 terminal cerci) comprising 40% of that segment's length .... *Hygrotus inaequalis*

   B Terminal projection of last segment comprising no more than 30% of that segment's length 39

39 A Tarsus of hind leg with row of spines along upper margin............................. *Coelambus* spp.

   B Tarsus of hind leg without row of spines along upper margin....................... *Hydroporus* spp.

40 A Abdominal segments 1–7 each with pair of long, filamentous, lateral gills, each gill as long or longer than abdominal width .... *Berosus* spp.

   B Abdominal segments without such gills 41

41 A Dorsal region of head with pair of parallel sutures extending to its posterior margin from the antennae .... *Laccobius biguttatus*

   B Sutures on dorsal region of head not parallel, but converge posteriorly and join at or before posterior margin 42

42 A Legs very short, not visible from above .... *Paracymus aeneus*

   B Legs short but visible in dorsal view.. 43

43 A Body ending in pair of long cerci; abdominal segments with chitinous plates dorsally .... *Helophorus* spp.

   B Body without long terminal cerci; abdominal segments without any chitinous plates............ 44

44 A Abdomen with prolegs on segments 3–7; 2nd thoracic segment with 1 dorsal chitinous plate; mandibles asymmetrical, left one with 2 teeth, right one with 1 .... *Enochrus* spp.

   B Without abdominal prolegs; 2nd thoracic segment with 2 dorsal chitinous plates; mandibles symmetrical .... *Hydrobius fuscipes*

Winged adults

Swimming, or more rarely burrowing, waterboatmen and beetles all conform to the standard body plan of their groups, with 3 pairs of legs of which 1 pair may be enlarged to form paddles, and hindwings that are hidden beneath hardened regions of the forewing (elytra in beetles; 'hemielytra' in the waterboatmen). Jansson (1986) describes European lesser waterboatmen and Friday (1988) describes British water beetles. See Fig. 89.

n.b. This key excludes the terrestrial insect fauna of salt-marshes, reed-beds and other fringing or emergent vegetation.

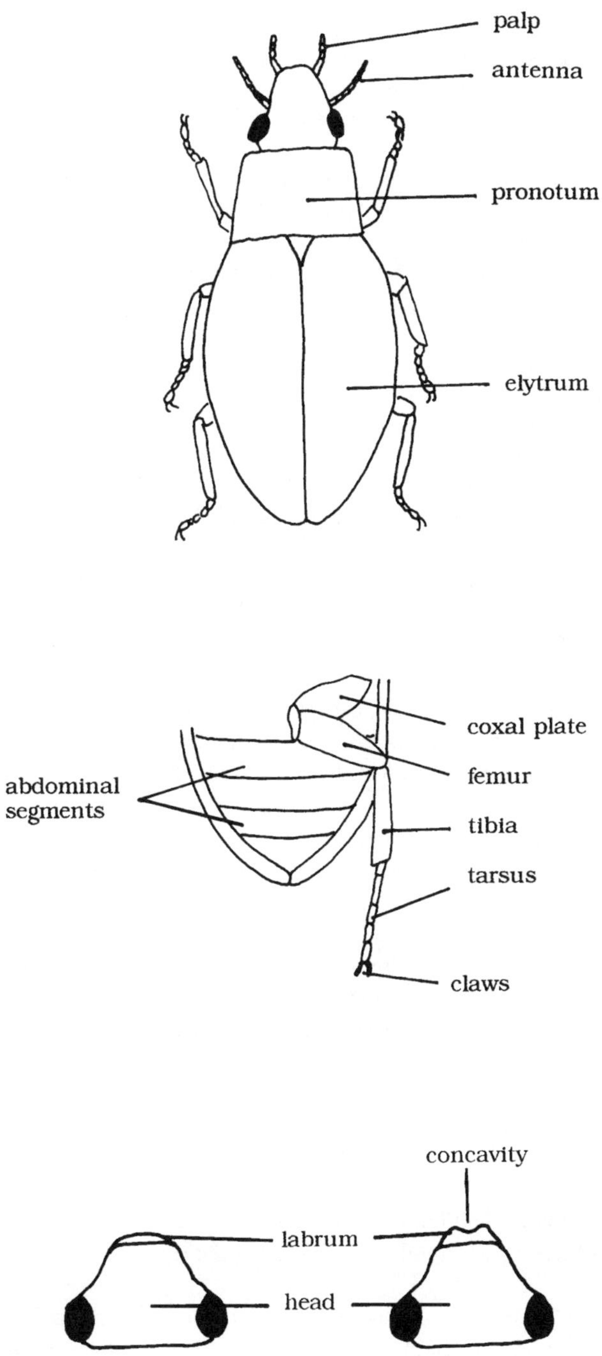

Fig. 89. Anatomical features used in the identification of adult insects.

1 A Animal on surface of intertidal mud, or in burrow within it... 2

B Animal associated with a body of water, including its surface.. 8

2 A Elongate, >4× as long as wide; wing cases minute, not covering abdomen............. *Bledius* spp.
[B. spectabilis is the only staphylinid recorded from brackish-water areas, but it is likely that others occur as well – many more are known from salt-marshes and marine beaches.]

B Body <3× as long as wide; wing cases cover abdomen... 3

3 A Hardened forewings overlap each other; head small, body forming a single smooth oval (thorax and abdomen with smooth lateral outline)...................................... *Saldula palustris*
[Other saldid bugs occur on salt-marshes and on shingle.]

B Wing cases (elytra) abut down dorsal midline of abdomen, but do not overlap; thorax and abdomen clearly differentiated... 4

4 A Body not densely hairy; tibiae of fore legs not enlarged; antennal articles not broader than long; pronotum broad anteriorly, tapering towards elytra; with projecting jaws.... carabid beetles
[Several carabids may venture temporarily onto estuarine mud, including from semi-aquatic salt-marshes, but the 4 mm long Cillenus laterale is a permanent inhabitant of estuarine mudflats.]

B Pronotum and elytra with dense hairs; tibiae of front legs very broad; antennal articles short and broader than long, the distalmost rounded.. 5

5 A Posterior face of pronotum not separated from dorsal surface by a raised ridge
.... *Heterocerus flexuosus*

B Posterior face of pronotum separated from dorsal surface by a raised ridge........................ 6

6 A Length 2.5–3.3 mm; antennae pale, except for last article which can be dark
.... *Heterocerus maritimus*

B Length 3.1–5.7 mm; antennae pale only at base if at all... 7

7 A Antennae completely dark; pronotum uniformly black .... *Heterocerus obsoletus*

B Basal part of antennae light coloured; pronotum with yellowish front margin and pale sides
.... *Heterocerus fenestratus*

8 A Wing cases (elytra) abut down dorsal midline of abdomen, but do not overlap; head with 7-11 articled antennae (that may be concealed in ventral pockets) and 3-articled palps........... 9

B Hardened forewings (hemielytra) overlap each other; head with antennae (may be very small) but not 3-articled palps 56

9 A Front legs clearly longer than middle and hind legs 10

B Front legs not longer than middle or hind legs.. 11

10 A Tarsal claws of middle and hind legs black, contrasting with the yellow legs....... *Gyrinus marinus*

B Tarsal claws of middle and hind legs yellow, like the legs .... *Gyrinus caspius*

11 A Antennae long (almost as long as body) and thread-like, inserted close together on the front of the head; tarsi of hind legs 4-articled (3rd small; 4th elongate); each elytrum with spine near tip .... *Macroplea mutica*
[Often recorded under the generic name Haemonia]

B Without this combination of characters 12

12 A Palps more obvious than antennae (which can be concealed in ventral pockets), from subequal to 4× antennal length; antennae with a club tip.. 13
   B Antennae not club-tipped 35

13 A Pronotum with 5 longitudinal furrows down its length.. 14
   B Pronotum without 5 such furrows 19

14 A One of the longitudinal rows of small pits on each elytrum not extending almost whole elytral length (the 2nd row from the midline)... 15
   B All rows of elytral pits running almost whole length of elytra 16

15 A Last article of palp symmetrical about its long axis whatever the viewpoint
   .... *Helophorus alternans*
   B Inner face of last article of palp distinctly less curved than outer face .... *Helophorus aequalis*

16 A Last article of palp symmetrical about its long axis whatever the viewpoint 17
   B Inner face of last article of palp distinctly less curved than outer face............................ 18

17 A Posterior part of lateral pronotal margins concave; last article of palp short (length 2× greatest width) .... *Helophorus arvernicus*
   B Posterior part of lateral pronotal margins straight; length of last article of palp >2.5× greatest width .... *Helophorus brevipalpis*

18 A Pronotum green to golden, front and side margins yellow .... *Helophorus minutus*
   B Pronotum dark brown, front and side margins brown .... *Helophorus flavipes*

19 A Pronotum with (a) median longitudinal groove, with or without 2 pairs of shallow pits on either side of it, or (b) 2 shallow transverse depressions one behind the other across the midline, or (c) deep transverse grooves right across it 20
   B Pronotum without large pits or grooves.. 28

20 A Pronotum with deep transverse grooves right across it and a median longitudinal groove
   .... *Ochthebius exaratus*
   B Pronotum without any transverse grooves extending right across it 21

21 A Pronotum without a median longitudinal groove 22
   B Pronotum with a median longitudinal groove... 24

22 A Elytra pale brown, paler than head .... *Ochthebius marinus*
   B Elytra very dark brown, not paler than head.. 23

23 A Length of body 1.7–2.0 mm; pronotum widest at level of anterior transverse depression
   .... *Ochthebius viridis*
   B Length of body 1.4–1.6 mm; pronotum widest at a level between the 2 transverse depressions .... *Ochthebius lenensis*

24 A Without paired pits on pronotum on either side of median longitudinal groove
   .... *Ochthebius aeneus*

B With 2 pits, one behind the other, on each side of the median longitudinal pronotal groove
.... 25

25 A Pronotum with scooped out posterolateral corner 26
B Pronotum narrower posteriorly than anteriorly but without scooped out corner 27

26 A Pronotum wide, its sides extending beyond shoulders of elytra; labrum with central concavity on front margin; length <1.9 mm... *Ochthebius auriculatus*
B Sides of pronotum not extending beyond shoulders of elytra; labrum with straight central region of front margin; length >1.8 mm .... *Ochthebius dilatatus*

27 A Length 2.5–2.8 mm.. *Ochthebius punctatus*
B Length 1.5–1.8 mm .... *Ochthebius nanus*

28 A Elytra with pits aligned in regular longitudinal rows 29
B Elytra with pits aligned randomly.. 32

29 A Each elytrum with 10 long rows of pits and 1 short row between 1st and 2nd rows............ 30
B Elytra without short incomplete rows of pits 31

30 A Each elytrum ending in a distinct spine; head yellow .... *Berosus spinosus*
B Elytra without terminal spines; head dark .... *Berosus affinis*

31 A Body >5 mm long; uniformly dark coloured, shiny..................................... *Hydrobius fuscipes*
B Body <3 mm long; elytra blotchy yellow and grey, pronotum with dark central area
.... *Laccobius biguttatus*

32 A Palps and antennae subequal in length; body 3 mm long; elytra with metallic sheen, legs and palps red .... *Paracymus aeneus*
B Palps much longer than antennae; body at least 5 mm long 33

33 A Head greenish-yellow or brown, without black marking............................... *Enochrus bicolor*
B Head with black region between eyes (if not more extensively)...................................... 34

34 A Pronotum and elytra yellow to orange; tip of palp darker than remainder
.... *Enochrus melanocephalus*
B Pronotum and elytra brown; palp uniformly pale coloured .... *Enochrus halophilus*

35 A Each elytrum with 10 longitudinal rows of pits; coxae of hind legs with large rounded plates extending over half ventral surface of abdomen and over proximal halves of the femora 36
[The genus Haliplus *keys out here. The species are very difficult to distinguish, and the features of the pronotum described below require a dry pronotal surface.]*
B Each elytrum with various grooves but without 10 longitudinal rows of pits; coxae of hind legs not greatly expanded.. 41

36 A Pronotum and coxal plates of hind legs covered with small pits (visible at ×20 magnification)... 37

B Pronotum and coxal plates of hind legs without such small pits 38

37 A Hind margin of pronotum with 2 short longitudinal furrows adjacent to 4th/5th row of elytral pits; elytra with continuous dark longitudinal lines .... *Haliplus confinis*
B Pronotum without such furrows; elytra with interrupted or blotchy dark longitudinal lines .... *Haliplus obliquus*

38 A Hind margin of pronotum without 2 short longitudinal furrows adjacent to 4th/5th row of elytral pits .... *Haliplus flavicollis*
B Hind margin of pronotum with such furrows 39
[The next 2 couplets require male beetles: those in which the tarsi of the fore legs bear tufts of pale hairs on the underside of the basal articles, and in which the basal article of the tarsi of the middle legs is swollen compared to the next article.]

39 A Claws of tarsus of front legs of equal length and curvature.......................... *Haliplus apicalis*
B One tarsal claw of front legs only some 2/3 length of other claw, and more strongly curved.. 40

40 A Basal article of tarsus of middle leg with its lower margin suddenly expanding distally .... *Haliplus immaculatus*
B Basal article of tarsus of middle leg smoothly expanding distally so that lower margin almost straight not sharply angled.. *Haliplus wehnckei*

41 A Body >25 mm length; pronotum with yellow borders, elytra metallic green... *Dytiscus circumflexus*
B Body <18 mm length 42

42 A Small triangular plate present between elytra anteriorly, at junction with pronotum; >6 mm body length 43
B Such a triangular plate absent; <6 mm body length 48

43 A Body >1 6mm long; elytra uniformly brown, pronotum darker but with pale borders .... *Colymbetes fuscus*
B Body <12 mm long 44

44 A Femur of hind leg without a series of 3–8 spines on its lower distal corner; pronotum yellow with diffuse dark markings along posterior margin; body 9–10 mm long..... *Rhantus exsoletus*
B Femur of hind leg with series of 3–8 spines on its lower distal corner; pronotal colouration not as above 45

45 A Head, pronotum and elytra uniformly dark coloured................................... 46
B Pronotum yellowish to mid-brown, much paler than (black) head 47

46 A Body >9.5 mm long; elytra uniformly dark .... *Agabus bipustulatus*
B Body up to 9 mm long; elytra with paler spots laterally and posteriorly .... *Agabus biguttatus*

47 A Pronotum with 2 black spots near midline; elytra olive green with black flecks; femora uniformly pale.. *Agabus nebulosus*
B Pronotum without black spots; elytra mottled brown; femora of front and middle legs with black patch.. *Agabus conspersus*

48 A Third article of tarsi of fore and middle legs not U-shaped; coxal processes of hind legs broad, combined breadth of processes > their length; central articles of antennae cup-shaped, with lobes in the male.. *Noterus clavicornis*

B Third article of tarsi of fore and middle legs U-shaped; coxal processes of hind legs longer than broad; central antennal articles not cup-shaped, and never lobed 49

49 A Body length >4 mm 50

B Body length <4 mm 52

50 A Posterior tip of each elytrum with a small tooth (view from ventral suface if necessary); sides of pronotum bulging, markedly convex...................................... *Potamonectes depressus*

B Posterior tip of each elytrum without any small tooth; sides of pronotum not bulging, ± straight 51

51 A Elytra chestnut with darker stripes .... *Coelambus impressopunctatus*

B Elytra yellow with black stripes.. *Coelambus parallelogrammus*

52 A Body markedly globose, strongly convex both dorsally and ventrally, body length only about 2× depth; head and anterior pronotum pale, elytra black with irregular chestnut markings .... *Hygrotus inaequalis*

B Body not globose; colour pattern not as above.. 53

53 A Anterior quarter of elytra pale and transparent, posterior three-quarters dark brown with pale flecks .... *Hydroporus tessellatus*

B Anterior quarter of elytra not transparent and not pale 54

54 A Head, pronotum and elytra a ± uniform reddish brown .... *Hydroporus angustatus*

B Head and pronotum black, with or without reddish markings 55

55 A Length 4.0–4.7 mm; central region of pronotum dull .... *Hydroporus planus*

B Length 3.2–3.8 mm; central region of pronotum shiny........................... *Hydroporus pubescens*

56 A Antennae long (much longer than head); head elongate, extending further anteriorly than eyes; middle legs longer than others; animal lives on water surface .... *Gerris thoracicus*

B Antennae inconspicuous; head broader than long, not markedly longer than eyes; middle legs not markedly longer than others; animal living in the water.................................... 57

57 A Mouthparts in form of long, pointed, ventral 'beak'; hind legs longer than others; animal swims upside-down 58

B Mouthparts in form of short, blunt, triangular flap; middle and hind legs subequal in length; animal doesn't swim upside-down 59

58 A Length <14.5 mm; anterior angles of pronotum pointed, embracing eyes; hemielytra with dark markings along anterior outer margin and at junction with the posterior membraneous section .... *Notonecta viridis*

B Length >14.5 mm; anterior angles of pronotum not pointed, not embracing eyes; hemiely-tra with dark markings only in form of irregular patches along anterior outer margin .... *Notonecta glauca*

59–68 Couplets 59–68 cover the lesser waterboatmen (Corixidae): often only the males of these can be identified to species. In males the tarsus of the fore leg (the 'pala') is flat and it bears 1–2 rows of stout pegs (the female palae are triangular in section and lack pegs); males also have an abdomen with asymmetrical segments (those of the females are symmetrical).

59 A Large, at least 3.5 mm broad, length 8–11 mm; in male, penultimate abdominal segment cleft on right-hand side 60

B Smaller, not more than 3 mm broad, length 5–9 mm; in male, penultimate abdominal segment cleft on left-hand side 61

60 A Length >10 mm; male pala with >30 pegs .... *Corixa panzeri*

B Length <9 mm; male pala with <30 pegs .... *Corixa affinis*

61 A Claw of hind legs entirely dark or with a dark spot proximally...................................... 62

B Claw of hind legs entirely pale 63

62 A Claw of hind legs entirely dark; male pala with pegs in a slightly curved line .... *Sigara lateralis*

B Claw of hind legs with a dark spot proximally, opposite a dark distal spot on tarsus; male pala with pegs in a sharply curved line .... *Paracorixa concinna*

63 A Male pala with pegs in a single row 64

B Male pala with pegs in 2 rows 66

64 A Male pala almost rectangular in shape, upper and lower margins subparallel, with pegs forming a right-angle bend distally; length >7 mm .... *Hesperocorixa sahlbergi*

B Male pala with straight lower and curved upper margin, its pegs not forming a distal right angle; length <6.5 mm 65

65 A Transverse ridge across front of male head at level of middle of eyes; upper margin of male pala with highest point of convexity half way along its length...................... *Sigara selecta*

B Transverse ridge across front of male head at level of lower edge of eyes; upper margin of male pala with highest point of convexity proximally, yielding a triangular shape... *Sigara stagnalis*

66 A Tarsus of hind legs with a dark distal mark .... *Callicorixa praeusta*

B Tarsus of hind legs entirely pale 67

67 A Proximal row of pegs of male pala in long, ± straight line at angle of 45° to the tarsal long axis; distal row very short .... *Sigara falleni*

B Proximal row of pegs of male pala shorter than distal row, aligned ± parallel to tarsal long axis.. 68

68 A Distal row of pegs of male pala with a sharp bend at extreme proximal end; body relatively pointed posteriorly .... *Sigara striata*

B Distal row of pegs of male pala without a sharp bend at extreme proximal end; body rounded posteriorly .... *Sigara dorsalis*

OLIGOENTOMATA
Collembola
Neanuridae

Anurida maritima (Fig. 90)

A small but obvious blue scavenger that occurs on salt-marsh mud and on adjacent areas of estuarine mudflats. It burrows beneath the surface for the duration of high tides. Up to 3 mm long.

ZYGOENTOMATA
Thysanura
Machilidae

Petrobius brevistylis (Fig. 90)

This bristle-tail can be found running across the substratum surface at high tidal levels all around the coast, whether the latter be rocky, shingly or composed of soft sediments. Up to 15 mm long.

PTERYGOTA
Ephemeroptera
Baetidae

Cloeon dipterum (Fig. 90)

Recorded sporadically from low salinity pools and lagoons. Up to 12 mm long.

Odonata
Coenagrionidae

Ischnura elegans (Fig. 90)

Although not restricted to brackish water, the nymphs of the blue-tail damselfly can develop in reedy pools, lagoons and even salt pans with salinities of over 12 and more rarely of over 18‰. Up to 25 mm long.

Libellulidae

Sympetrum sanguineum (ruddy darter) (Fig. 90) and *Orthetrum cancellatum* (black-tailed skimmer) (Fig. 90) Dragonfly nymphs have rarely been recorded from brackish water, although *O. cancellatum* (up to 26 mm long) frequently occurs in the Dybso fjord, Denmark, at salinities of up to 13‰, and elsewhere in less saline pools, and *S. sanguineum* (up to 17 mm long) has been recorded from a salinity of 8‰ in a brackish pool in Hampshire, England.

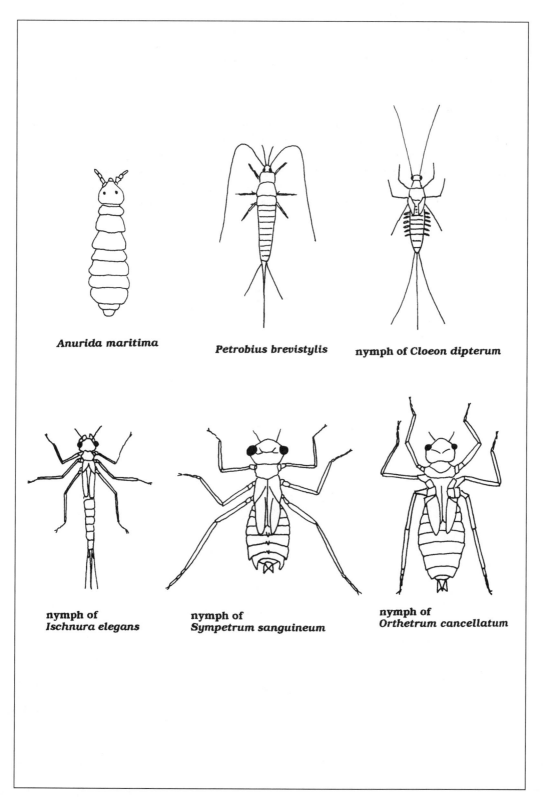

Anurida maritima

Petrobius brevistylis

nymph of Cloeon dipterum

**nymph of
Ischnura elegans**

**nymph of
Sympetrum sanguineum**

**nymph of
Orthetrum cancellatum**

Fig. 90. Brackish-water insects (I).

Hemiptera
Corixidae

Paracorixa concinna (Fig. 91); *Callicorixa praeusta* (Fig. 91); *Sigara stagnalis* (Fig. 91); *S. selecta* (Fig. 91); *S. dorsalis* (Fig. 91); *S. lateralis* (Fig. 91); *S. striata* (Fig. 91); *S. falleni* (Fig. 91); *Hesperocorixa sahlbergi* (Fig. 92); *Corixa affinis* (Fig. 92); *C. panzeri* (Fig. 92)

Corixids are omnivorous insects taking algae, detritus and some animal food. A few species appear specialist – *S. falleni* may solely consume algae, *C. panzeri* solely animals, and *S. stagnalis* and *S. lateralis* only detritus, for example – although as yet there is insufficient evidence to be sure of differences between the various species. All characterize weedy pools, ditches and lagoons, with abundant *Potamogeton, Ruppia* or *Chaetomorpha,* and have wide European distributions. None are restricted to brackish waters, although *S. selecta* and *S. stagnalis* come closest to this and probably have optima in the vicinity of 6–8‰, and maxima of some 20‰. Salt-marsh pools with higher salinities can be colonized temporarily.

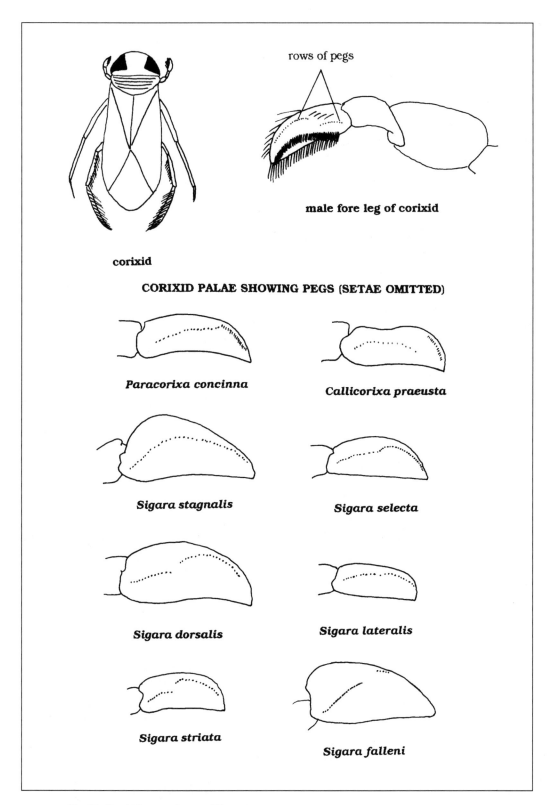

rows of pegs

male fore leg of corixid

corixid

CORIXID PALAE SHOWING PEGS (SETAE OMITTED)

Paracorixa concinna

Callicorixa praeusta

Sigara stagnalis

Sigara selecta

Sigara dorsalis

Sigara lateralis

Sigara striata

Sigara falleni

Fig. 91. Brackish-water insects (II).

Gerridae

Gerris thoracicus (Fig. 92)

Recorded occasionally from the surface of salt pans, brackish pools and lagoons, where they feed on organisms trapped in or falling on the surface water film. It also occurs in brackish water in the northern Baltic. Up to 12 mm long.

Notonectidae

Notonecta glauca (Fig. 92) and *N. viridis* (Fig. 92)

These two predatory greater waterboatmen, which feed mainly on insect larvae and small corixids, occur in relatively low salinity brackish pools and lagoons. Up to 16 mm long.

Saldidae

Saldula palustris (Fig. 92)

A fast-running predatory inhabitant of the surface of salt marshes and estuarine mudflats that can, at least at the nymph stage, withstand temporary submersion by tidal water and can even remain active when covered. Up to 5 mm long.

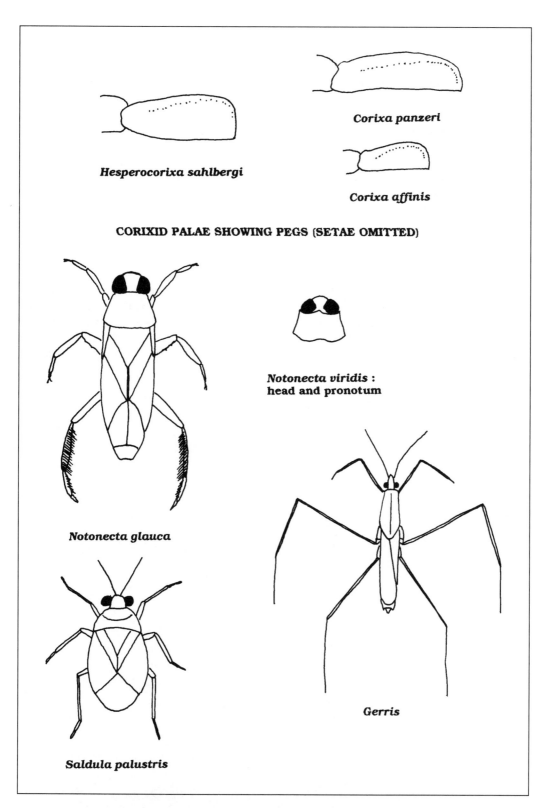

Fig. 92. Brackish-water insects (III).

Trichoptera
Limnephilidae
Limnephilus affinis (Fig. 93)
Although not restricted to brackish water, the larvae of this species can develop in pools, lagoons and salt pans with salinities of up to nearly 20‰ and can survive in waters of over 25‰. In an inhabited salt pan in Iceland, the salinity range was recorded as 22–35‰. They appear to be omnivorous, consuming detritus and dipteran larvae. Up to 18 mm long.

Leptoceridae
Mystacides longicornis (Fig. 93) and *Oecetis ochracea* (Fig. 93)
The larvae of these essentially freshwater species have on occasion been recorded in pools with salinities of up to 10‰. Up to 14 mm long.

Lepidoptera
Pyralidae
Acentropus niveus (Fig. 93)
The caterpillars and wingless females of the 'false caddis-fly moth' have frequently been recorded, in densities of up to 150 individuals/m², amongst *Potamogeton, Ruppia* and *Zostera* in lagoons and similar habitats, in waters of up to 15‰, and temporarily 25‰, salinity. The young larval stages mine the leaves of these plants. Up to 12 mm long.

Nymphula nympheata (Fig. 93)
Caterpillars of the brown china mark moth have been recorded from *Potamogeton* in a low-salinity lagoon in Cornwall, England. Up to 12 mm long.

larva of *Limnephilus*

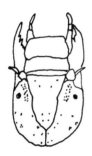

***Mystacides longicornis* larva in case**

**head of larva of
*Oecetis ochracea***

larva of *Nymphula nympheata*

**protective larval case of
*Acentropus niveus***

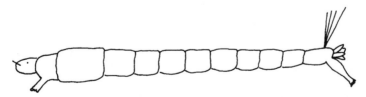

larva of chironomid

Fig. 93. Brackish-water insects (IV).

Diptera

Chironomidae (Fig. 93); Dolichopodidae (Fig. 94); Psychodidae (Fig. 94); Ptychopteridae (Fig. 94); Stratiomyiidae (Fig. 94); Syrphidae (Fig. 94); Tabanidae (Fig. 94); Tipulidae (Fig. 94); Ephydridae (Fig. 94)

Fly larvae, especially chironomids, dolichopodids and tipulids, are abundant in estuarine mud throughout the whole salinity range (including in full-strength sea water). Chironomids are also an important component of the weed fauna and sediments in tideless brackish waters, whilst stratiomyiids and ephydrids (particularly *Ephydra riparia*) occur amongst *Ruppia*, *Chaetomorpha* and floating algal mats in such habitats. *E. riparia* clasp the *Ruppia*, etc. during pupation. Adult dolichopodid flies may even be significant predators on the aquatic fauna, having been recorded as capturing and consuming *Corophium*!

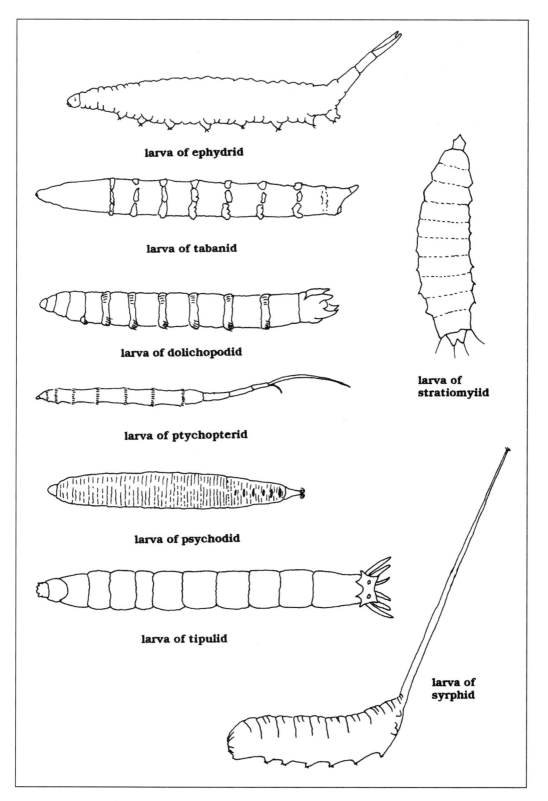

larva of ephydrid

larva of tabanid

larva of dolichopodid

larva of ptychopterid

larva of psychodid

larva of tipulid

larva of
stratiomyiid

larva of
syrphid

Fig. 94. Brackish-water insects (V).

Coleoptera
Gyrinidae (whirligigs)

Gyrinus caspius (Fig. 95) and *G. marinus* (Fig. 95)

Whirligig adults swim along the air/water interface of tideless brackish-water habitats, there feeding on living or dead animals trapped in the surface film. Each eye is divided into two portions, one for underwater and the other for aerial vision. The rather centipede-like larvae are probably carnivorous. Of the two species, *G. marinus* is in spite of its name less-frequently encountered in brackish water (and more in fresh water) than *G. caspius*. Up to 8 mm long.

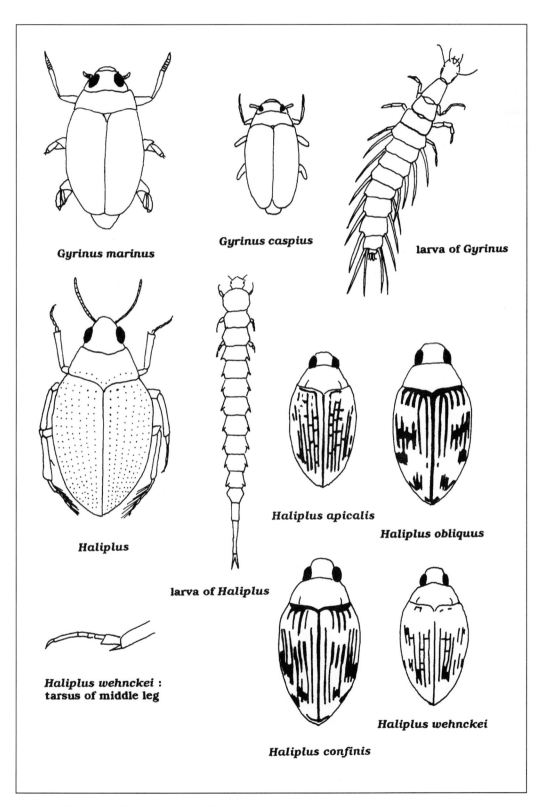

Gyrinus marinus

Gyrinus caspius

larva of Gyrinus

Haliplus

larva of Haliplus

Haliplus apicalis

Haliplus obliquus

Haliplus wehnckei :
tarsus of middle leg

Haliplus confinis

Haliplus wehnckei

Fig. 95. Brackish-water insects (VI).

Haliplidae

Haliplus wehnckei (Fig. 95); *H. apicalis* (Fig. 95); *H. obliquus* (Fig. 95); *H. confinis* (Fig. 95); *H. immaculatus* (Fig. 96); *H. flavicollis* (Fig. 96)

Haliplus species, which are probably herbivorous at all stages of their life history, feeding on fine algae, occur in weedy brackish pools, ditches and lagoons with salinities of less than some 10‰, although *H. apicalis* can withstand more concentrated media; pupation occurs out of water. Some of the air store is located beneath the large coxal plates that typify the family. Adults are good swimmers, the larvae somewhat sluggish. Up to 5 mm long.

Noteridae

Noterus clavicornis (Fig. 96)

Like *Macroplea* below, the larva of *Noterus* taps into the air passages of submerged portions of vegetation, in this case by burrowing into the bottom sediments and piercing the bases of the stems of *Phragmites* using a spine on the tip of its abdomen, and it also pupates there within a cocoon. The aquatic adult is a predator, but what it consumes in brackish water is unknown. Recorded from relatively low-salinity lagoons and pools in southwestern England and the Netherlands. Up to 5 mm long.

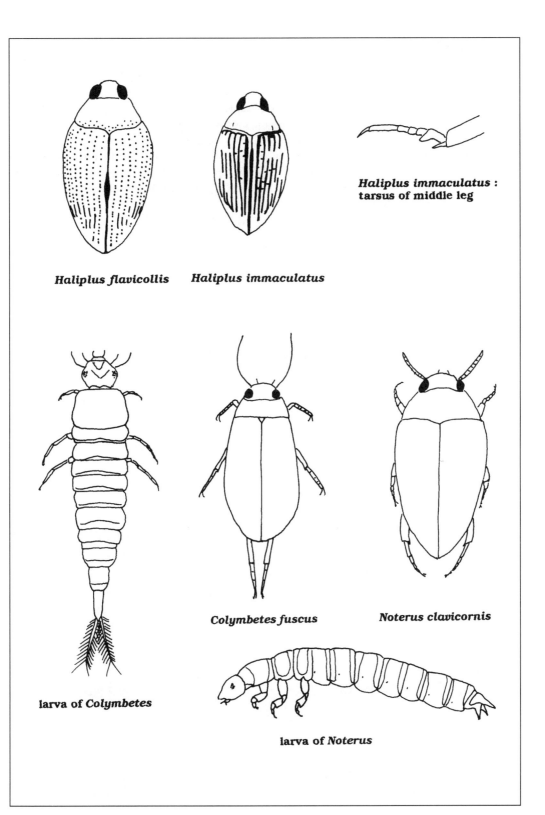

**Haliplus immaculatus :
tarsus of middle leg**

Haliplus flavicollis   **Haliplus immaculatus**

Colymbetes fuscus   **Noterus clavicornis**

larva of Colymbetes

larva of Noterus

Fig. 96. Brackish-water insects (VII).

Dytiscidae

Colymbetes fuscus (Fig. 96); *Rhantus exsoletus* (Fig. 97); *Agabus biguttatus* (Fig. 97); *A. conspersus* (Fig. 97); *A. bipustulatus* (Fig. 97); *A. nebulosus* (Fig. 97); *Dytiscus circumflexus* (Fig. 98); *Hygrotus inaequalis* (Fig. 98); *Coelambus impressopunctatus* (Fig. 98); *C. parallelogrammus* (Fig. 98); *Hydroporus planus* (Fig. 99); *H. angustatus* (Fig. 99); *H. pubescens* (Fig. 99); *H. tessellatus* (Fig. 99); *Potamonectes depressus* (Fig. 99)

Like the hydrophilids, dytiscids or diving beetles are common in weedy brackish pools, ditches and lagoons. The larvae are carnivores, as are many of the adults, although some of the latter are probably scavengers: their brackish-water diets are unknown. Adults, and most larvae, exchange respiratory gases by breaking the surface-water film with the tip of the abdomen. Most dytiscids seem restricted to waters of less than 10‰ salinity, although *Colymbetes* can inhabit waters twice as concentrated. Most are good swimmers, and the larvae are also relatively active.

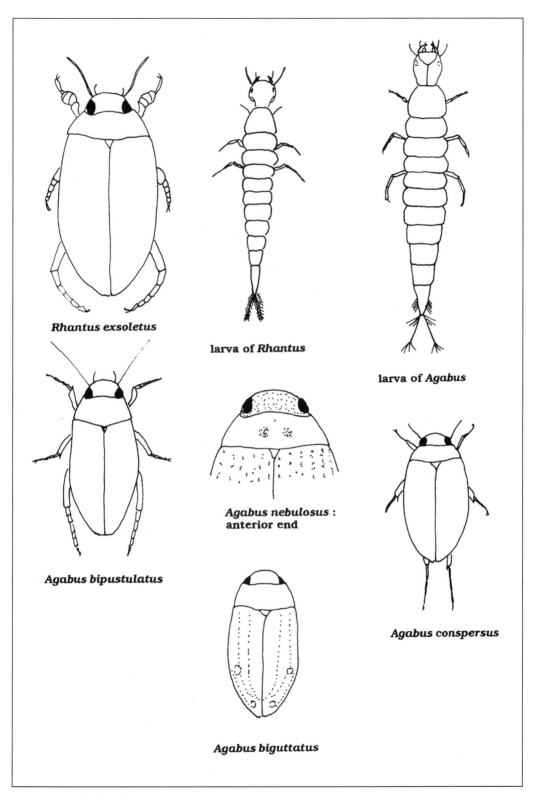

Rhantus exsoletus

larva of Rhantus

larva of Agabus

Agabus bipustulatus

**Agabus nebulosus :
anterior end**

Agabus conspersus

Agabus biguttatus

Fig. 97. Brackish-water insects (VIII).

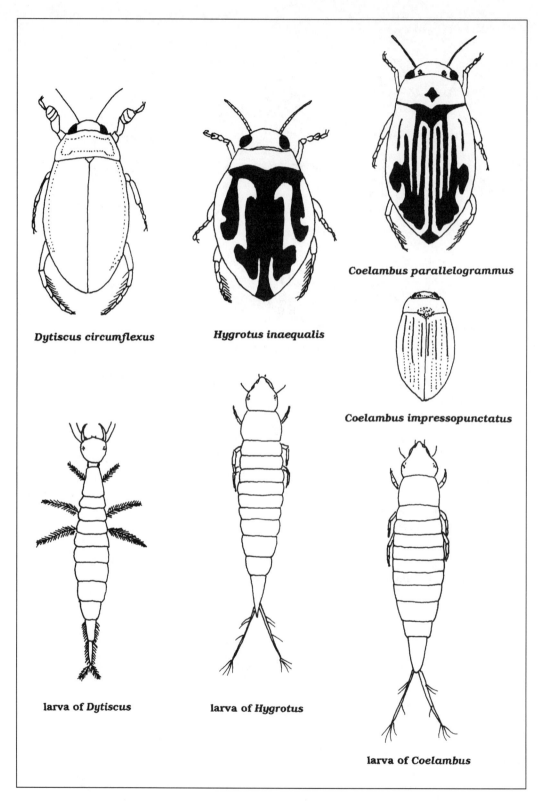

Dytiscus circumflexus

Hygrotus inaequalis

Coelambus parallelogrammus

Coelambus impressopunctatus

larva of Dytiscus

larva of Hygrotus

larva of Coelambus

Fig. 98. Brackish-water insects (IX).

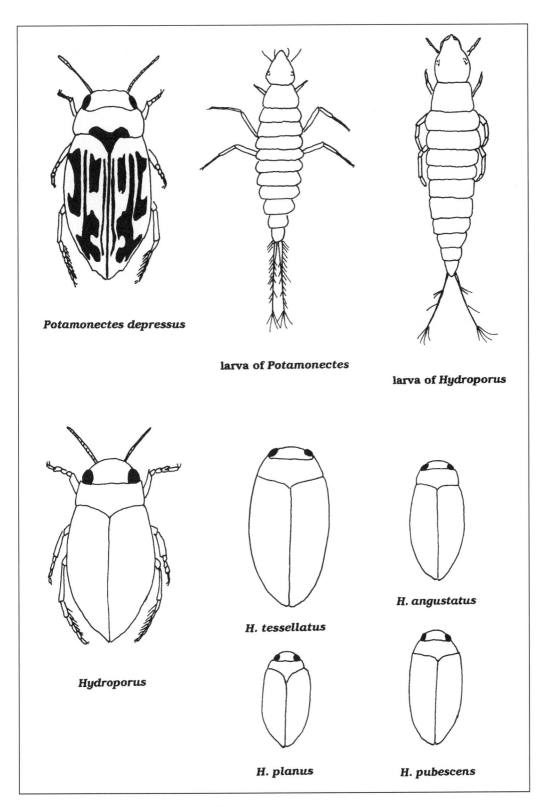

Fig. 99. Brackish-water insects (X).

Hydrophilidae

Helophorus brevipalpis (Fig. 100); *H. aequalis* (Fig. 100); *H. minutus* (Fig. 100); *H. flavipes* (Fig. 100); *H. arvernicus* (Fig. 100); *H. alternans* (Fig. 100); *Berosus affinis* (Fig. 100); *B. spinosus* (Fig. 100); *Enochrus bicolor* (Fig. 101); *E. halophilus* (Fig. 101); *E. melanocephalus* (Fig. 101); *Laccobius biguttatus* (Fig. 101); *Paracymus aeneus* (Fig. 101); *Hydrobius fuscipes* (Fig. 102)

The hydrophilids or scavenger beetles are a large group of water beetles. The larvae are mostly short-lived, slow-moving and predatory, the adults are omnivorous. Unusually amongst water beetles, the adults break the surface film to exchange gases not with the tip of the abdomen but with the side of the head. The larvae of *Berosus* have gills, but the larvae of the other genera are air-breathers and those of *Helophorus* may be entirely terrestrial. Brackish-water hydrophilids typically occur in weedy ditches, pools and lagoons, especially those with *Ruppia* or *Potamogeton*, although *Berosus spinosus* and *Enochrus bicolor* can also occur in estuarine and fully saline (even hypersaline) conditions, and *Helophorus* species may be found in salt pans. Only *Berosus* swims, the other adults crawl. Very little is known of their ecology, and the larvae of some species are unknown. In Mediterranean lagoons, at least, the rare *Paracymus* inhabits beds of submerged vegetation and rarely occurs in the open water.

Hydraenidae

Ochthebius marinus (Fig. 102); *O. punctatus* (Fig. 102); *O. dilatatus* (Fig. 102); *O. viridis* (Fig. 102); *O. lenensis* (Fig. 102); *O. nanus* (Fig. 102); *O. exaratus* (Fig. 102); *O. aeneus* (Fig. 102); *O. auriculatus* (Fig. 102)

Hydraenids are closely related to the hydrophilids and share many features with them. The larvae are all terrestrial, however, and the small adults (only 1–3 mm long) characteristically occur in leaf litter and thick vegetation where they probably feed on detritus. Brackish pools, ditches, small lagoons and salt pans are colonized throughout the whole brackish (and marine) salinity range. (Adult *O. quadricollis*, not present in northwestern Europe, can occur in salinities of 270‰, and complete their development at 94‰.)

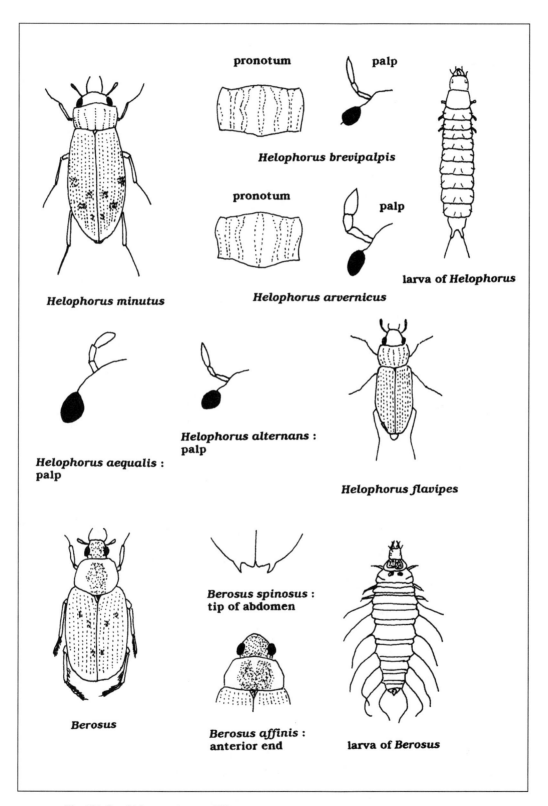

Fig. 100. Brackish-water insects (XI).

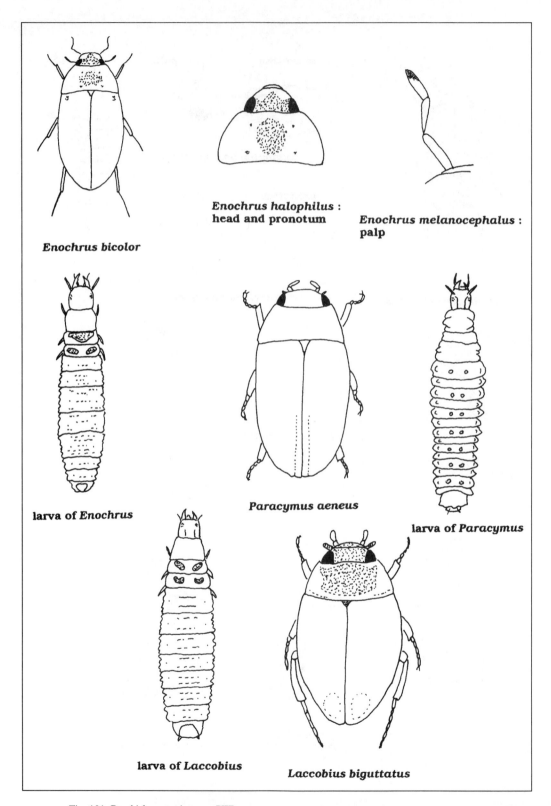

Enochrus bicolor

Enochrus halophilus :
head and pronotum

Enochrus melanocephalus :
palp

larva of Enochrus

Paracymus aeneus

larva of Paracymus

larva of Laccobius

Laccobius biguttatus

Fig. 101. Brackish-water insects (XII).

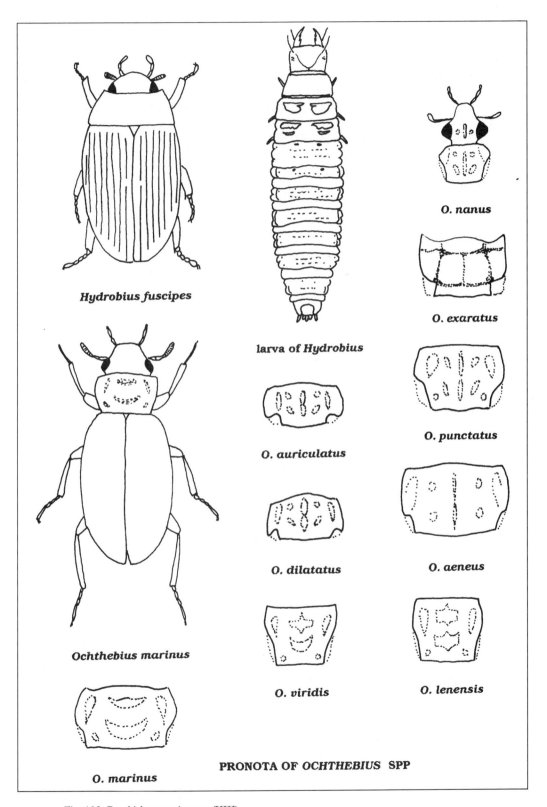

Hydrobius fuscipes

larva of Hydrobius

O. nanus

O. exaratus

O. auriculatus

O. punctatus

O. dilatatus

O. aeneus

Ochthebius marinus

O. viridis

O. lenensis

O. marinus

PRONOTA OF OCHTHEBIUS SPP

Fig. 102. Brackish-water insects (XIII).

Chrysomelidae
Macroplea mutica (Fig. 103)

The long-lived larvae of this species, widely recorded under the generic name *Haemonia*, tap into the air passages of the roots/rhizomes of *Ruppia, Zostera* or *Potamogeton* in lagoons and equivalent habitats, and hence can remain below the water surface feeding on the plant tissues; the pupa occurs in the same microhabitat in a silken cocoon. The adults, which are also aquatic, have been described as connecting their antennae to air bubbles on the surface of submerged leaves to channel supplies of oxygen. It can achieve densities of over 500 individuals/m², and can tolerate salinities of up to 20‰. Up to 8 mm long.

Staphylinidae
Bledius spp. (Fig. 103)

Several species of *Bledius* burrow into coastal muds and sands from the high-water mark down to below high-water neap. They are not aquatic in that all stages are air-breathing, either obtaining air during low tide or from stores maintained within the burrow systems. Algae probably provide the main source of food. *B. spectabilis,* which commonly occurs in estuarine muds, is perhaps the most thoroughly intertidal species. Up to 8 mm long.

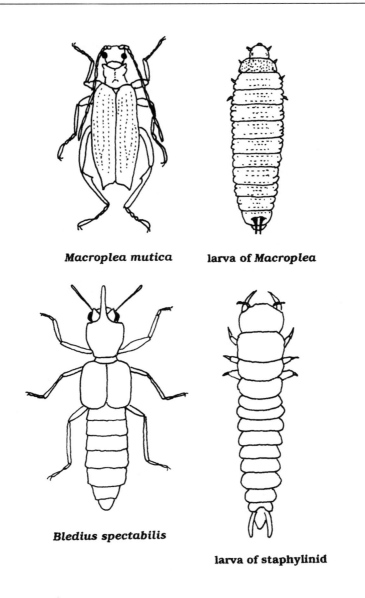

Macroplea mutica   **larva of Macroplea**

Bledius spectabilis

larva of staphylinid

larva of Heterocerus

Fig. 103. Brackish-water insects (XIV).

Heteroceridae

Heterocerus flexuosus (Fig. 104); *H. obsoletus* (Fig. 104); *H. fenestratus* (Fig. 104); *H. maritimus* (Fig. 104)
Like *Bledius* above, species of *Heterocerus* burrow in intertidal estuarine muds but are not really aquatic, being active only at low tide or within the confines of air-filled burrows. They too probably largely subsist on algae.

Carabidae

Cillenus laterale (Fig. 104)
Cillenus, more familiar under the generic name *Bembidion*, is a surface-dwelling predator of estuarine mudflats, feeding *inter alia* on *Corophium* and dolichopodid larvae. Its larvae live in air-filled burrows within the sediment; the adults roam over the surface during low tide. Up to 5 mm long.

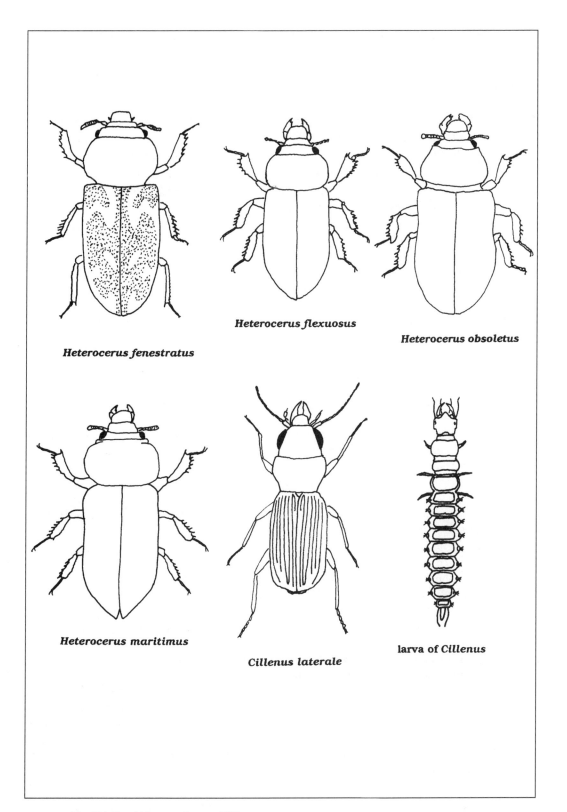

Heterocerus flexuosus

Heterocerus obsoletus

Heterocerus fenestratus

Heterocerus maritimus

Cillenus laterale

larva of Cillenus

Fig. 104. Brackish-water insects (XV).

Bryozoans (moss animals or sea-mats)

Colonial animals encrusting stones or vegetation that in brackish waters occur in the form of (a) thousands of minute, calcareous, rectangular boxes (each 0.3–0.8 mm long and 0.2–0.4 mm broad, and each containing one individual bryozoan, known as a zooid) in sheets or mats, or (b) more diffuse, open colonies of 0.5–2 mm long zooids arising individually or in clumps from a network of creeping stolons, or, more rarely, (c) groups of zooids (0.5–0.6 × 0.3–0.4 mm) set in a communal gelatinous matrix. Bryozoans of type 'b' may appear superficially similar to hydroids, although there is no danger of confusing any of the brackish-water species with a hydroid. Ryland & Hayward (1977), Hayward & Ryland (1979), Mundy (1980) and Hayward (1985) describe species likely to occur in brackish water. See Fig. 105.

1 A Colony calcified and therefore hard and brittle... 2
   B Colony not calcified: gelatinous, rubbery, horny, etc... 11

2 A Colony an encrusting sheet sometimes rising from the surface in the form of foliose plates.... 3
   B Colony erect and bush-like, attached to the substratum by a single 'stalk'.............. *Bugula* spp.

3 A Part of the surface of the individual zooids membraneous, so that the zooid within is partly visible... 4
   B Whole of the surface of the individual zooids (apart from the aperture) calcified................ 10

4 A With one or more short or long spines projecting from the proximal and/or lateral margins of the membraneous frontal surface... 5
   B Without any spines arising from the proximal or lateral margins of the membraneous frontal surface... 9

5 A Each or some zooids with 1–2 swollen, rounded or triangular chambers (containing small zooids) near the aperture 6
   B Without such chambers 7

6 A Operculum crescentic; accessory zooids triangular, without a pointed lid-like projection; zooids >0.4 mm long... *Conopeum reticulum*
   B Operculum semicircular; accessory zooids swollen and rounded, with a pointed, triangular lid-like projection; zooids <0.4 mm long.. *Callopora aurita*

7 A Calcified area of frontal surface with numerous pores....................................... *Electra pilosa*
   B Calcified area of frontal surface without pores 8

8 A Operculum opaque (because calcified), visible as semicircular structure with concave lower edge; without spines on lateral margins of membraneous zone, with single stout proximal spine... *Electra crustulenta*
   B Operculum not calcified; often with at least one pair of lateral spines besides the slender, curved proximal one... *Electra monostachys*

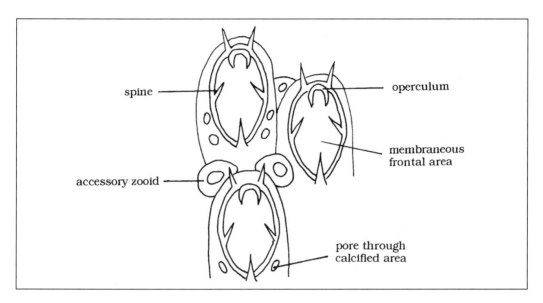

Fig. 105. Anatomical features used in the identification of bryozoans.

9 A With two triangular chambers at the apertural end of each zooid .... *Conopeum reticulum*
  B With or without triangular chambers at the apertural end of each zooid, but if present, not in pairs.. *Conopeum seurati*

10 A Frontal surface coarse, covered with ridges and pits; aperture large and bell-shaped .... *Cryptosula pallasiana*
  B Frontal surface smooth; aperture small, shaped as in Fig. 108................. *Turbicellepora avicularis*

11 A Colony forming a thin, gelatinous sheet of zooids; individual zooids flat, irregular in outline .... *Alcyonidium gelatinosum*
  B Colony not forming a thin, gelatinous sheet of zooids; individual zooids cylindrical 12

12 A Colony a dense mound of contiguous tubes with a honeycomb-like surface appearance .... *Plumatella fungosa*
  B Colony not in the form of a mound of dense, contiguous tubes.................................... 13

13 A Colony with zooids in fused groups of 10–16 (5–8 pairs in a double row) separated by intervening 'internodes' on a dichotomously branching wiry stolon................ *Amathia lendigera*
  B Colony not like this.. 14

14 A Zooid with numerous feeding tentacles (up to 70) arranged in a horseshoe-shape .... *Plumatella repens*
  B Zooid with few feeding tentacles (<15) arranged in a circle 15

15 A Colony comprising a series of zooids clearly issuing from a single branching stolon 16
  B Colony of zooids not clearly issuing from a single stolon, but connected together by extensions from their bases; often budding new zooids from the bodies of older ones .... *Victorella pavida*

16 **A** Zooids elongate, club-shaped, their bases narrowing into stalks; with 11 or more feeding
tentacles .... *Farrella repens*

    B Zooids short, arising directly from the stolon; with 10 or fewer feeding tentacles 17

17 **A** Stolon markedly thinner than the zooids; zooids usually issuing singly, in pairs or in small
groups; colony never forming erect tufts; zooid with 8 tentacles .... *Bowerbankia gracilis*

    B Stolon thicker than the bases of the zooids; zooids issuing in dense clusters; colony creep-
ing or forming bush-like tufts; zooid with 10 tentacles......................... *Bowerbankia imbricata*

PHYLACTOLAEMATA
Plumatellida
Plumatellidae

Plumatella repens (Fig. 106)

An essentially freshwater species that throughout the region (and in the Mediterranean) can extend
its range into dilute brackish waters in non-tidal lagoons with salinities of less than 10‰. It grows
over submerged vegetation and stones, and in at least one lagoon can share these with *Victorella*.

Plumatella fungosa (Fig. 106)

An essentially freshwater species that can extend its range into dilute brackish waters in non-tidal
ponds and lagoons with salinities of less than 8‰. It can grow into large mounds (with diameters in
metres) attached to pilings, the submerged parts of trees, etc. Within northwest Europe its known
brackish-water stations are mostly along the southern shores of the North Sea.

GYMNOLAEMATA
Ctenostomata
Alcyonidiidae

Alcyonidium gelatinosum (Fig. 106)

This gelatinous bryozoan forms a thin, smooth, colourless sheet on algae in estuaries and it may also
be the species, recorded as *A. polyoum* – a synonym of *A. gelatinosum*, that encrusts gastropod shells
in The Fleet, Dorset. (The large, erect, rubbery species that used to be termed *A. gelatinosum* should
now be known as *A. diaphanum*.)

Victorellidae

Victorella pavida (Fig. 106)

Known from a few scattered lagoonal habitats extending from Brittany and Cornwall to The
Netherlands and the Baltic, and elsewhere from the Mediterranean and Atlantic, Indian and Pacific
Oceans. It occurs on stones, wood, vegetation (*Ruppia, Phragmites,* etc.), and bivalve shells in salinities down to, and locally including, fresh water and often in regions experiencing considerable fluctuations in salinity (from seawater to freshwater coverage during different parts of a single day). It is
also known from estuarine docks (it was first described from the London Docks). It regresses to
hibernacula during cold winters or periods of summer drought.

Triticellidae

Farrella repens (Fig. 106)

F. repens is known from scattered estuaries from the Solway Firth around the west and south coast
of England to the southern North Sea (Essex round to Denmark). It encrusts stones, shells,
hydroids, and other bryozoans.

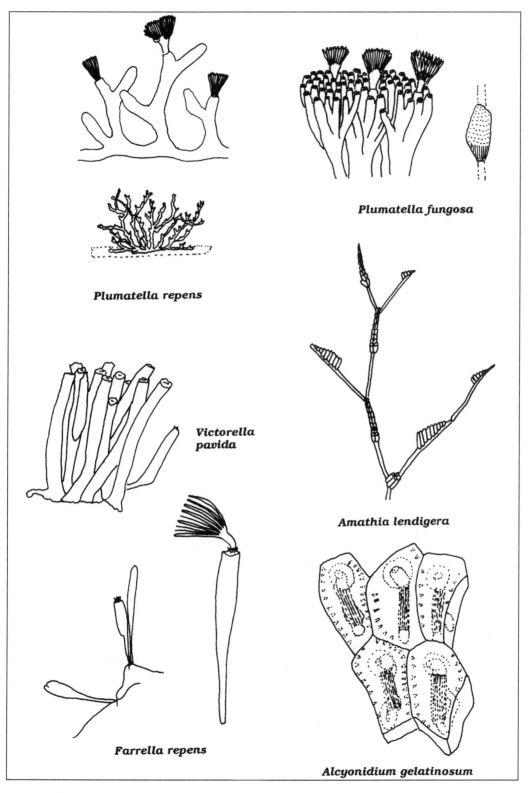

Fig. 106. Brackish-water bryozoans (I).

Vesiculariidae
Amathia lendigera (Fig. 106)
This diffuse species occurs, tangled around hydroid and bryozoan colonies or attached to shells, in the west of the region and in the Channel. It can colonise the more saline regions of estuaries.

Bowerbankia imbricata (Fig. 107)
A common species throughout the region, *B. imbricata* is known from coastal lagoons around the North Sea and Channel shores, and in estuaries attached to fucoid algae (particularly *Fucus vesiclosus* and *Ascophyllum nodosum*).

Bowerbankia gracilis (Fig. 107)
B. gracilis is a well known, cosmopolitan, brackish-water fouling organism throughout the region (and into the Baltic). It grows over man-made structures in docks and harbours, as well as on stones, shells, submerged vegetation (e.g. *Ruppia*) and colonial hydroids or bryozoans in estuaries (down to salinities of some 15‰) and lagoons (down to some 5‰). It is sometimes recorded under the name *B. caudata* .

Cheilostomata
Membraniporidae
Conopeum seurati (Fig. 107)
Perhaps the most common encrusting bryozoan in northwest European lagoons. It grows on wooden posts, stones, emergent (*Phragmites*) and submerged (*Ruppia*) vegetation in salinities of right down to 1‰. It occurs around the North Sea, the Channel and, more rarely (although this may simply reflect the relative scarcity of lagoons), in the west of the region, besides being found in the Baltic and Mediterranean.

Conopeum reticulum (Fig. 107)
C. reticulum appears to be the estuarine counterpart of *C. seurati* above, occurring, widespreadly and often abundantly, in salinities of down to some 18‰, as well as in the sea, over the same geographical range as the latter species. It is largely restricted to hard substrata.

Electridae
Electra crustulenta (Fig. 107)
A common encrusting brackish-water species, often recorded under the generic name *Membranipora*, that occurs down to 2‰ in the Baltic and to some 10‰ elsewhere. It has been widely recorded from the southern half of the region, in both estuaries and lagoons, growing on stones, concrete, *Ruppia* and fucoid algae. In a few lagoons and tidal flats it forms masses deserving the name reefs.

Electra pilosa (Fig. 107)
This cosmopolitan species is abundant in northwest European coastal seas, and it occasionally penetrates into brackish waters of over 20‰ salinity, being especially common on estuarine fucoid algae.

Electra monostachys (Fig. 107)
An estuarine and coastal species, *E. monostachys* encrusts shells and stones in the mouths of estuaries throughout the southern half of the region.

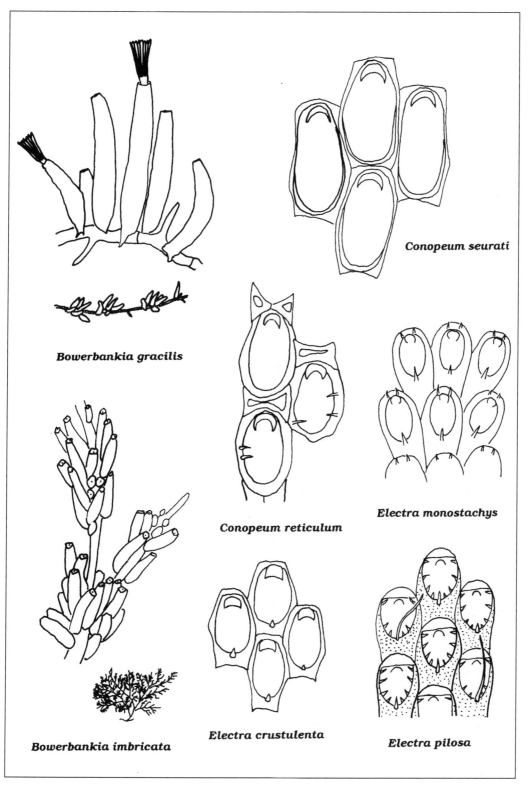

Conopeum seurati

Bowerbankia gracilis

Conopeum reticulum

Electra monostachys

Bowerbankia imbricata

Electra crustulenta

Electra pilosa

Fig. 107. Brackish-water bryozoans (II).

Calloporidae
Callopora aurita (Fig. 108)
Often recorded as *Membranipora aurita*, this species occurs throughout the region (though common-est in the north) on shells and stones. Although it penetrates part way into the Baltic it is relatively rare in brackish water outside that sea. It is, however, present in the mouths of some estuaries.

Bugulidae
Bugula spp. (Fig. 108)
Several species of *Bugula* are common fouling organisms in ports and harbours, including those in the mouths of estuaries, and in areas heated by power-station or other thermal discharges. They form tufts or small bushes attached to buoys, pilings, boats (including their water intake pipes), and other firm surfaces. The colonies die back in the autumn and overwinter as stolons or special peren-nating zooids.

Cryptosulidae
Cryptosula pallasiana (Fig. 108)
A common intertidal, rocky-shore and not otherwise brackish-water species that is present in the Widewater lagoon, Sussex.

Celleporidae
Turbicellepora avicularis (Fig. 108)
A widely distributed species, that encrusts shells, hydroids and other bryozoans throughout the region, which has occasionally been recorded from the mouths of estuaries.

Echinoderms

Ophiuroids (brittlestars)

Flat echinoderms with their body in the form of a small, central, more-or-less circular disc from which issue five long, narrow, sinuous arms. The only species likely to be encountered in brackish waters is *Amphipholis squamata*, with a disc diameter of <5 mm, which may be common in high salinity lagoons. Mortensen (1927) describes the British echinoderms.

OPHIUROIDEA
Amphiuridae
Amphipholis squamata (Fig. 108)
A common and cosmopolitan species that occurs under stones, and occasionally on soft sediments, in shallow marine waters. It is also known in English shingle lagoons with salinities of >20‰, from East Anglia round to Hampshire.

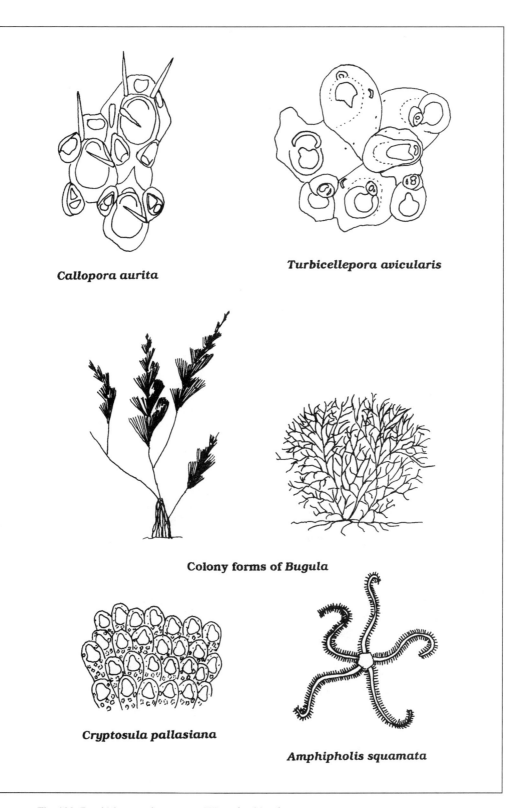

Callopora aurita

Turbicellepora avicularis

Colony forms of Bugula

Cryptosula pallasiana

Amphipholis squamata

Fig. 108. Brackish-water bryozoans (III) and echinoderms.

Ascidians (tunicates or sea-squirts)

Solitary or colonial animals attached to or encrusting firm surfaces (stones, vegetation, etc.). Colonial species are jelly-like and often brightly coloured, with the individual 'zooids' arranged in stars or in zip-fastener patterns; solitary species are encased in tough, wrinkled sacs with two openings (siphons) from which water squirts if the sacs are gently squeezed. Millar (1970) describes British species.

1 A Colonial, the individual zooids being arranged in numerous star-shaped clusters all set in a communal jelly; highly colourful – green, yellow, blue, red, brown – the zooids contrasting with the colour of the jelly.. *Botryllus schlosseri*
  B Non-colonial 2

2 A Jelly grey to greenish, spherical, up to 30 mm diameter, attached directly to the substratum .... *Molgula manhattensis*
  B Jelly elongate, attached to the substratum by a stalk; total height up to 12 cm .... *Styela clava*

TUNICATA
Pleurogona
Styelidae
Botryllus schlosseri (Fig. 109)
A common encrusting or pendant species that occurs from Norway throughout the region to the Mediterranean. It is abundant in parts of The Fleet, Dorset, but does not otherwise appear to be notably brackish-water in its habits.

Styela clava (Fig. 109)
An accidentally introduced species (from Japan or Korea) that has established itself in the Channel, from Hampshire to Devon, and along the south Welsh coast. It is abundant at several sites, for example in Southampton Water, where it lives attached to stones and dock substrata. It has been recorded from one Channel coast lagoon.

Molgulidae
Molgula manhattensis (Fig. 109)
A widely distributed North Atlantic species that attaches to a variety of hard substrata, including stones, pilings and algae in harbours and in the mouths of estuaries throughout the region. It has been recorded once from a Channel coast lagoon.

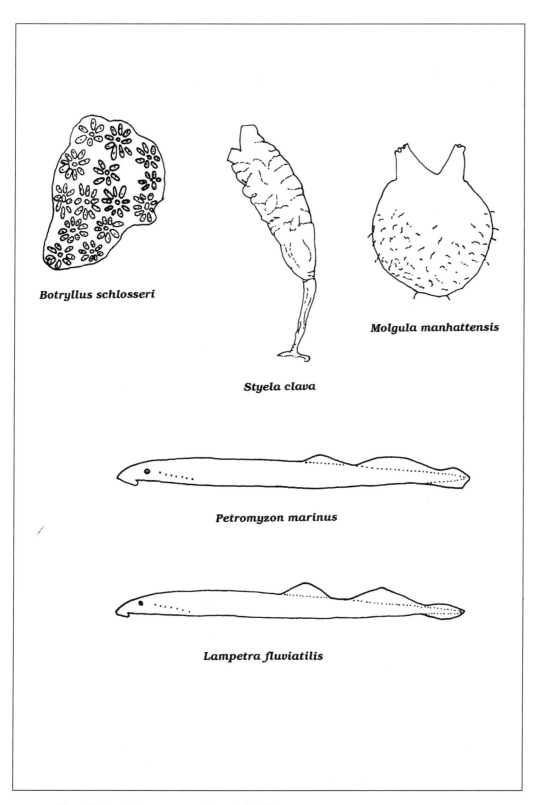

Fig. 109. Brackish-water sea-squirts and fish (I).

FISH

Elongate, sometimes extremely elongate, animals with a distinct head at one end bearing a mouth and a pair of eyes; fins are present at least along parts of the dorsal and ventral midlines and around the tail end (see Fig. 110). The body may be covered by scales, bony plates or naked skin. Wheeler (1969; 1978) and Bagenal (1973) describe all the fish of the region.

1 **A** Extremely elongate fish, with lengths >12× greatest body height 2
    B Fish with lengths <10× greatest body height 9

2 **A** Mouth in form of round or oval ventral sucker; with 7 pairs of gill openings; without paired fins but with dorsal and tail fins 3
    B Mouth not forming ventral sucker; with 1 pair gill openings; with paired fins or, if absent, without a tail fin either 4

3 **A** Adults with black mottling dorsally; up to 1 m long; mouth with many circles of teeth .... *Petromyzon marinus*
    B Grey; up to 50 cm long; mouth with 1 circle of teeth .... *Lampetra fluviatilis*

4 **A** Needle-like jaws projecting in the form of a beak, the lower jaw extending further than the upper; with small dorsal fin just in front of caudal fin .... *Belone bellone*
    B Jaws not projecting; dorsal fin not as above 5

5 **A** Dorsal fin elongate, merging with caudal fin; scales minute .... *Anguilla anguilla*
    B Dorsal fin small, half-way along dorsal margin; scales large and bony................................. 6

6 **A** Without pectoral or caudal fins .... *Nerophis ophidion*
    B Pectoral and caudal fins present... 7

7 **A** Snout laterally compressed, > half height of head .... *Siphonostoma typhle*
    B Snout cylindrical, < half height of head 8

8 **A** With >16 rings of scales between head and dorsal fin; snout > half length of head; up to 50 cm length.. *Syngnathus acus*
    B With <16 rings of scales between head and dorsal fin; snout < half length of head; up to 17 cm length... *Syngnathus rostellatus*

9 **A** Flat fish that lie on their side (n.b. both eyes are on the same side of the head), with widths >40% of their length; body oval or diamond-shaped with 'lateral' fringe of fins 10
    B Body not as above.. 16

10 **A** Front of head smoothly rounded; mouth not terminal 11
    B Front of head pointed; mouth terminal with prominent lower jaw 12

11 **A** Pectoral fin on blind side of body only slightly smaller than that on eyed side; up to 50 cm long... *Solea solea*

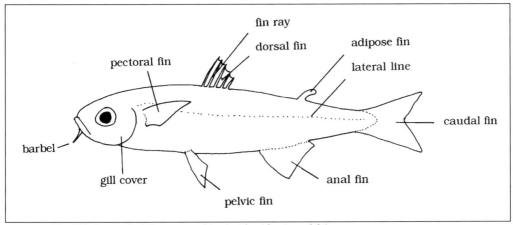

Fig. 110. Anatomical features used in the identification of fish.

B Pectoral fin on blind side of body reduced to 1 long ray + 1–2 small rays; every 4th–7th 'lateral' fin ray black; <15 cm long .... *Buglossidium luteum*

12 A Dorsal fin extends to point in front of both eyes; eyes on left side of body 13
   B Dorsal fin does not extend in front of eyes; eyes on right side of body 14

13 A Upper surface without scales, but with few, large, bony spines .... *Scophthalmus maximus*
   B Upper surface with small scales, without bony spines .... *Scophthalmus rhombus*

14 A Lateral line with marked curve over pectoral fin .... *Limanda limanda*
   B Lateral line not curved over pectoral fin 15

15 A Upper surface smooth, with vivid red or orange spots; anal fin with >47 rays
   .... *Pleuronectes platessa*
   B Upper surface rough, with dull orange or no spots; anal fin with <47 rays .... *Platichthys flesus*
   [*These 2 spp. can hybridize, hybrids being intermediate in characteristics.* Platichthys *can also occur as mirror-image individuals, with their eyes on the left-hand side of the body.*]

16 A With isolated movable spines in front of the dorsal fin 17
   B Without isolated movable spines in front of dorsal fin 19

17 A With <5 (usually 2 large and 1 small) dorsal spines .... *Gasterosteus aculeatus*
   B With >7 equisized dorsal spines 18

18 A With 8–10 dorsal spines .... *Pungitius pungitius*
   B With 15 dorsal spines .... *Spinachia spinachia*

19 A Pelvic fins located posteriorly to pectoral fins 20
   B Pelvic fins located anteriorly to, or immediately below, pectoral fins 33

20 A With 2 dorsal fins containing fin rays 21
   B With only 1 rayed dorsal fin (an adipose fin may be present dorsally as well) 24

21 A 1st dorsal fin with >4 rays; body with iridescent silver line along each side; jaw oblique .... *Atherina presbyter*

    B 1st dorsal fin with 4 rays; body without intense silver line, although dark longitudinal stripes may occur; jaw short not oblique 22

22 A *Adults* : Upper lip thick (> half eye diameter) and with papillae along lower margin. *Juveniles* (25–50 mm long) : Ventral region of head in front of eyes with black pigment-cells over whole area .... *Crenimugil labrosus*

    B *Adults* : Upper lip thin (< half eye diameter), without papillae. *Juveniles* (25–50 mm long) : Ventral region of head in front of eyes with longitudinal zones devoid of black pigment-cells 23

23 A *Adults* : Pectoral fin short, only just reaching eye (at most) when folded forwards; greyish blue dorsally, silvery laterally and ventrally. *Juveniles* (25–50 mm long) : Black pigment cells present along ventral midline of head and laterally, but with pigmentless zone between the two .... *Liza ramada*

    B *Adults* : Pectoral fin elongate, reaching beyond posterior margin of eye when folded forwards; colour as above but with golden hue on the sides and golden spots on side of head. *Juveniles* (25–50 mm long) : Black pigment cells present laterally on ventral part of head, but not along ventral midline .... *Liza auratus*

24 A With a dorsal adipose fin.. 25

    B Without an adipose fin 28

25 A Mouth extends posterior to the eye; dorsal fin begins at or behind base of pelvic fins; smells of cucumber.. *Osmerus eperlanus*

    B Mouth extends only as far as eye; dorsal fin begins in front of pelvics; no smell of cucumber 26

26 A With feeble, almost toothless jaws; with fleshy, pointed snout................... *Coregonus oxyrinchus*

    B Jaws not feeble or almost toothless; without fleshy, pointed snout................................. 27

27 A Dorsal fin with 10–12 rays; external upper jaw bone not extending beyond posterior margin of eye; when fish <15 cm long, with deeply forked and sharply pointed caudal fin, with <3 dark spots on gill cover, and, if line of large dark patches present along side of body, then with 10-12 of such patches; when fish >15 cm long, caudal fin shallowly forked .... *Salmo salar*

    B Dorsal fin with 8–10 rays; external upper jaw bone extending well beyond posterior margin of eye; when fish <15 cm long, with shallowly forked and bluntly ending caudal fin, with >3 black spots on gill cover, and, if line of large dark patches present along side of body, then with 9 or 10 such patches; when fish >15 cm long, caudal fin unforked.......... *Salmo trutta*

28 A With pointed snout; jaws extending posteriorly to the eye .... *Engraulis encrasicolus*

    B Without pointed snout; jaws not extending posterior to the eye 29

29 A Gill cover smooth 30

    B Gill cover with radial ridges.. 31

30 A Scales along ventral midline with sharp, backwardly directed points (spiny to touch); dorsal fin begins behind base of pelvics; with green dorsal colouring........................ *Sprattus sprattus*

    B Scales along ventral midline not sharply pointed (although still rough to the touch); dorsal fin begins in front of, or directly above, base of pelvics; with blue dorsal colouring
.... *Clupea harengus*

31 A With large lateral scales (*c.* 30 from head to tail); without any dark lateral spots; without notch in midline of upper jaw .... *Sardina pilchardus*

    B With small scales (*c.* 60–80 from head to tail); with 1 or more dark lateral spots; with notch in midline of upper jaw 32

32 A With 6–10 dark, round, lateral spots; with <70 scales between head and tail .... *Alosa fallax*

    B With 1–5 dark lateral spots; with >70 scales from head to tail .... *Alosa alosa*

33 A Eel-like with slimy skin; dorsal, caudal and anal fins form one continuous long, low fin
.... *Zoarces viviparus*

    B Dorsal, caudal and anal fins not confluent... 34

34 A Head with 2 dorsal and 1 ventral barbels .... *Rhinonemus cimbrius*

    B Head with or without 1 ventral barbel, but without dorsal barbels................................. 35

35 A With 3 dorsal fins, 2 anal fins, and 1 ventral barbel; with black spot at base of pectoral fin
.... *Trisopterus luscus*

    B With 2 dorsal and 1 anal fins; without a barbel 36

36 A Pelvic fins fused together... 37

    B Pelvic fins separate 39

37 A Region between end of base of 2nd dorsal fin and start of caudal fin short, < half length of base of 2nd dorsal fin; up to 17 cm long .... *Gobius niger*

    B Region between end of base of 2nd dorsal fin and start of caudal fin elongate, as long or longer than base of 2nd dorsal fin; <10 cm long.. 38

38 A With triangular dark mark on upper part of base of pectoral fin; 1st dorsal fin without black spot posteriorly .... *Pomatoschistus microps*

    B Without dark mark on base of pectoral fin; 1st dorsal fin with white-rimmed black spot near its rear edge .... *Pomatoschistus minutus*

39 A Body with scales; gill cover without spines; pectoral fins small; up to 1m long
.... *Dicentrarchus labrax*

    B Body without scales, with or without bony plates; gill cover with spine/s; pectoral fins large and spiny; up to 30 cm long... 40

40 A Body surface covered by bony plates; with sharp spines on tip of snout; region between end of base of 2nd dorsal fin and start of caudal fin elongate, longer than base of 2nd dorsal fin... *Agonus cataphractus*

    B Body surface naked; without spines on tip of snout; region between end of base of 2nd dorsal fin and start of caudal fin short, much shorter than base of 2nd dorsal fin
.... *Acanthocottus scorpius*

AGNATHA
Cyclostomata
Petromyzonidae (lampreys)
Petromyzon marinus (sea lamprey) (Fig. 109)
Sea lampreys breed and spend their larval life (as 'prides' or ammocoetes) in fresh water, but their adulthood in the sea. Young lampreys, of some 15–20 cm length, therefore pass through estuaries on their way to the sea (usually in winter), and adults, which die after spawning, undertake the reverse migration (in autumn or early winter). The fish is not common, however, and is rarely seen or caught in brackish water.

Lampetra fluviatilis (lampern) (Fig. 109)
The brackish-water biology of the lampern is effectively the same as that of the sea lamprey above.

OSTEICHTHYES
Isospondyli
Clupeidae
Clupea harengus (herring) (Fig. 111) and *Sprattus sprattus* (sprat) (Fig. 111)
Herring (up to 40 cm long) and sprats (up to 14 cm long) are coastal, pelagic marine fish that can withstand considerable dilution of sea water, occurring for example in the Baltic. The juveniles are especially coastal and occur in considerable numbers in estuaries, feeding on copepods and other small crustaceans. These are harvested commercially and marketed as 'whitebait'.

Sardina pilchardus (pilchard or sardine) (Fig. 111)
The pilchard is less tolerant of low-salinity water than the sprat or herring, but young fish may nevertheless occur in the mouths of estuaries, especially in the south and west of the region. Up to 25 cm long.

Engraulis encrasicolus (anchovy) (Fig. 111)
Anchovies are pelagic marine fish occurring in the south of the region (from the southern Irish and North Seas southwards) that enter estuarine areas, especially those of The Netherlands, to spawn in the summer months. Up to 20 cm long.

Alosa alosa (allis shad) (Fig. 111) and *Alosa fallax* (twaite shad) (Fig. 111)
These marine fish enter rivers to spawn, the allis shad journeying much farther into fresh water to do so than the twaite. Adults pass through estuaries in spring, spawn and then return to sea; young allis shad spending 1–2 years in fresh water before migrating to sea, the juvenile twaite entering the estuaries during their first autumn. Up to 60 cm long.

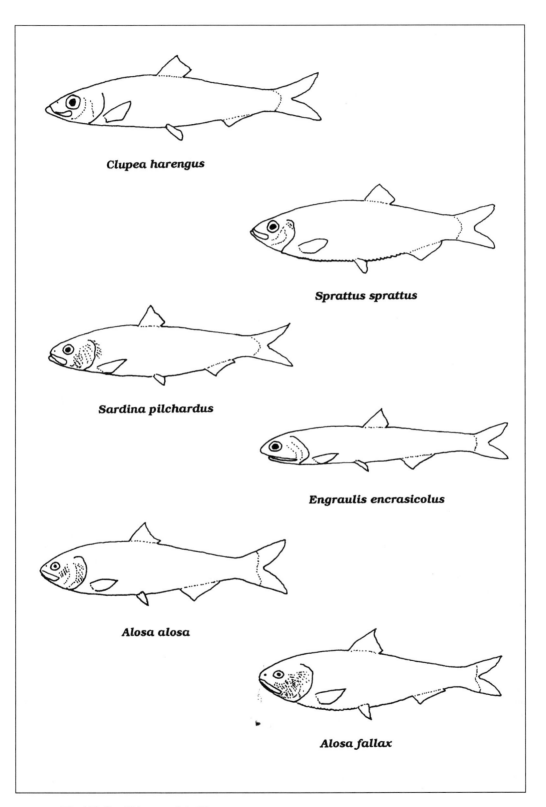

Fig. 111. Brackish-water fish (II).

Coregonidae

Coregonus oxyrinchus (houting) (Fig. 112)

The houting used to be a brackish-water fish of the southern North Sea and the Baltic. The Baltic forms still survive, but the southern North Sea population is now probably extinct. It inhabited estuarine regions from the Rhine to Denmark, spawning in the adjacent rivers, but rarely occurring in the North Sea itself. Up to 50 cm long.

Salmonidae

Salmo salar (salmon) (Fig. 112) and *S. trutta* (sea trout) (Fig. 112)

These salmonids spawn in fresh water, spend most of their adult lives at sea, and therefore pass through estuaries at least twice during their life cycles. Adults migrate into the rivers in late summer/early autumn, and many die after spawning. Survivors (some moribund) return to sea in late autumn/winter. The young spend one or more years (up to 5) in fresh water and eventually migrate into estuaries when some 20 cm long, acclimate to higher-salinity water for a time, and then move out to sea. Adult sea trout are more coastal than adult salmon, and many probably feed within estuaries (on amphipods, mysids and small decapods) during high tide.

Osmeridae

Osmerus eperlanus (smelt) (Fig. 112)

The smelt is an up to 20 cm long, estuarine and coastal fish of the southern North Sea (and Baltic) that preys on crustaceans (amphipods and isopods) and other small fish (e.g. gobies and 'whitebait'). Adults spawn in estuaries or the lower reaches of rivers in spring, the eggs being demersal, and the young remain strictly estuarine for their first summer. The adults may remain in brackish waters for the rest of their lives or may move somewhat offshore, to return again in winter preparatory to the spring breeding season.

Apodes

Anguillidae

Anguilla anguilla (eel) (Fig. 112)

Eels first enter estuaries and lagoons connected to the sea as 7 cm long, transparent elvers ('glass eels') between January and April (depending on area). Many pass straight through the brackish zones and enter fresh waters, eventually to return on the seawards spawning migration when some 6 or 7 years old (and 30–50 cm (males) or 40–100 cm (females) long). Whilst in fresh waters the eels are yellowish ventrally and grey, brown or black dorsally ('yellow eels'), but when migrating to the sea in the autumn the colour changes to silver ventrally and becomes darker dorsally ('silver eels'). Many yellow eels, however, rather than being freshwater, remain in in lagoons and estuaries, and inhabit muddy sediments, in lagoons beneath *Zostera* or *Chaetmorpha*, and there prey on crabs, sticklebacks, *Nereis* and amphipods. Adults die after spawning.

Synentognathi

Belonidae

Belone bellone (garfish) (Fig. 112)

The garfish is a pelagic species that occurs mainly in open water but which enters shallow coastal waters occasionally in summer and when young, and which may be seen in the mouths of estuaries. Up to 75 cm long.

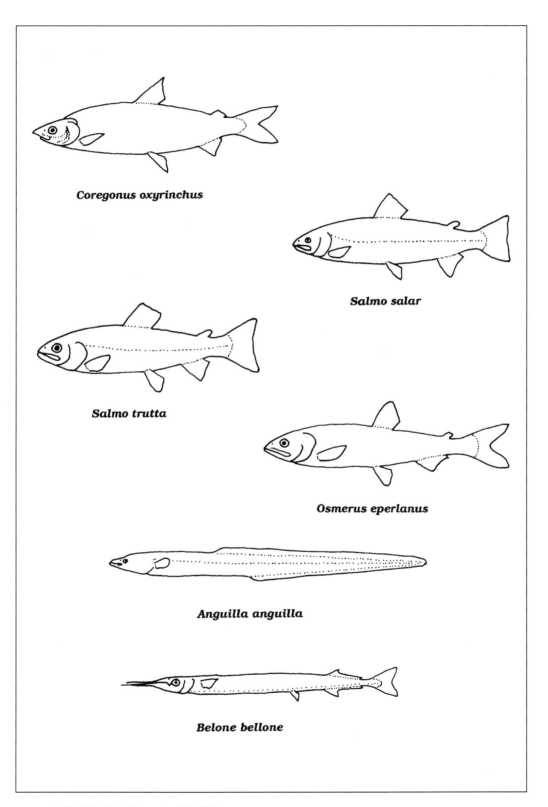

Fig. 112. Brackish-water fish (III).

Solenichthyes
Syngnathidae

Siphonostoma typhle (broad-nosed pipefish) (Fig. 113), *Syngnathus acus* (great pipefish) (Fig. 113), *Syngnathus rostellatus* (lesser pipefish) (Fig. 113) and *Nerophis ophidion* (worm pipefish) (Fig. 113)

Pipefish are associated with *Zostera* and other submerged vegetation throughout the region, including within estuaries and in those lagoons that have an open connection to the sea. The extent to which they can withstand brackish water is, however, poorly known. The diet includes the young of such fish as gobies, and amphipods, isopods and similarly sized crustaceans.

Anacanthini
Gadidae

Trisopterus luscus (bib or pout) (Fig. 113)

The bib is an inshore gadoid that is common throughout the region. The young stages in particular occur in estuaries during the summer months, where they feed on shrimps, prawns and *Carcinus*. Up to 40 cm long.

Rhinonemus cimbrius (4-bearded rockling) (Fig. 113)

This is mainly a deep-water species, but the occasional individual may be found in the mouth of an estuary, especially in the north of the region and in the winter months. Up to 40 cm long.

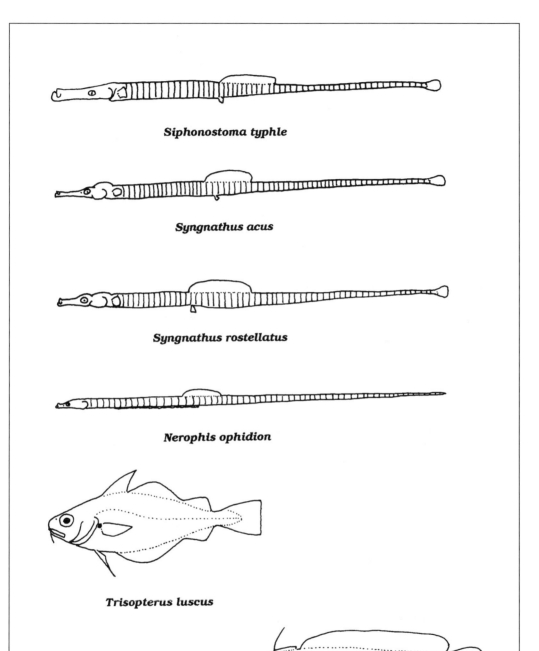

Siphonostoma typhle

Syngnathus acus

Syngnathus rostellatus

Nerophis ophidion

Trisopterus luscus

Rhinonemus cimbrius

Fig. 113. Brackish-water fish (IV).

Percomorphi
Serranidae
Dicentrarchus labrax (bass) (Fig. 114)
Bass are summer visitors to estuaries in the southwestern half of the region, where they often congregate around hot-water discharges, as from coastal power stations. Spawning occurs offshore, but
young bass (3–6 cm long) are commonly found in estuaries in the autumn. When in estuaries the
diet includes crustaceans (amphipods, isopods, shrimps) and young fish. Up to 1 m long.

Gobiidae
Gobius niger (black goby) (Fig. 114)
This typically brackish-water fish is relatively seldom seen because it only occurs in deep water areas
(>2 m) of estuaries (and the Baltic). It frequents areas of soft sediment where it eats anything that it
can catch, from *Idotea* to *Cerastoderma, Arenicola* and young flatfish.

Pomatoschistus microps (common goby) (Fig. 114)
P. microps is perhaps the most characteristic resident brackish-water fish in northwest Europe. It is
found in all types of brackish water including totally land-locked lagoons, and it has been recorded
from salinities of down to 4‰. Although often abundant in areas with dense beds of submerged
vegetation, the goby appears to prefer the open water areas over expanses of bare muddy or sandy
sediment. Females lay their eggs there on the underside of empty *Mya* or *Cerastoderma* shells partly
projecting from the sediment; the eggs then being guarded by the male. Where this is possible, there
is a migration into deeper waters for the duration of the winter. The diet includes amphipods,
isopods, oligochaetes and chironomid larvae when adult, and interstitial copepods, etc. when young.
Up to 7 cm long.

Pomatoschistus minutus (common goby) (Fig. 114)
In comparison with *P. microps* above, this species occurs in deeper waters (>2 m) and in higher-
salinity regions, although the breeding biology and the diet of the two species are very similar.
P. minutus is found in estuaries throughout the region, but not generally in lagoons. Up to 10 cm
long.

Zoarcidae
Zoarces viviparus (eelpout) (Fig. 114)
The eelpout is a northern fish that occurs around the shores of the North Sea (and Baltic) and
reaches its southern limit in the eastern Channel. It occurs in estuaries (mainly in the south) and
estuarine lagoons, especially in *Zostera* meadows, where it consumes *Idotea* and other crustaceans,
gastropod and bivalve molluscs, and small fish. An alternative name is the viviparous blenny: fertilization occurs in the ovary and the young are born as miniature adults. Up to 45 cm long.

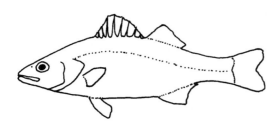

Dicentrarchus labrax

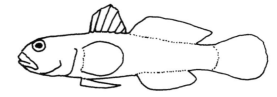

Gobius niger

Pomatoschistus minutus

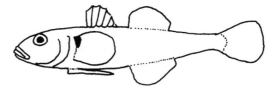

Pomatoschistus microps

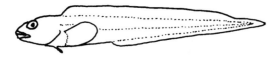

Zoarces viviparus

Fig. 114. Brackish-water fish (V).

Mugilidae

Crenimugil labrosus (thick-lipped grey mullet) (Fig. 115)

Young grey mullet 20–50 mm long, as well as fully-grown adults (up to 70 cm long), commonly enter estuaries and lagoons with an open connection to the sea in the southwestern half of the region in spring and summer. There they graze micro-algae, together with interstitial and surface-dwelling animals, from the sediment surface. In winter they may hibernate in warmer, deeper water. The 20 mm long juveniles usually arrive in July and August.

Liza ramada (thin-lipped grey mullet) (Fig. 115) and *Liza auratus* (golden mullet) (Fig. 115)

Juvenile (20–50 mm long) and fully-grown adults (up to 70 cm long) of these two grey mullets less commonly enter estuaries and lagoons with an open connection to the sea in the southwestern half of the region than does the thick-lipped species. Their brackish-water biology, however, is the same as their more common relative. The 20 mm long juveniles are thought to arrive between February and April. For further information on juvenile grey mullets, see Reay & Cornell (1988).

Atherinidae

Atherina presbyter (sand-smelt) (Fig. 115)

Like the grey mullet to which it is related, the small sand-smelt is commonest in the southwestern half of the region, although it does also occur along the southern shores of the North Sea, and it is also a summer visitor to estuaries, open lagoons, harbours, etc. They spawn as well as feed in brackish waters, their food mainly comprising small crustaceans. Up to 15 cm long.

Scleroparei

Cottidae

Acanthocottus scorpius (sea scorpion) (Fig. 115)

Also known under the generic name *Myoxocephalus* and the common name 'father lasher', this is a species reaching its southern limit in the Bristol Channel and along the southern shores of the North Sea. It is common in the marine halves of estuaries amongst algae and over bare sediment, feeding on crustaceans and young fish. If caught, the fish should be handled with care since cuts made by its spines can turn septic. Up to 30 cm long.

Agonidae

Agonus cataphractus (bullhead or pogge) (Fig. 115)

The bullhead has the same range as the sea scorpion above. In the south it appears in the mouths of estuaries during the winter months, although in the north it can be found in such habitats in the summer as well. Its diet appears largely to be small crustaceans. Up to 15 cm long.

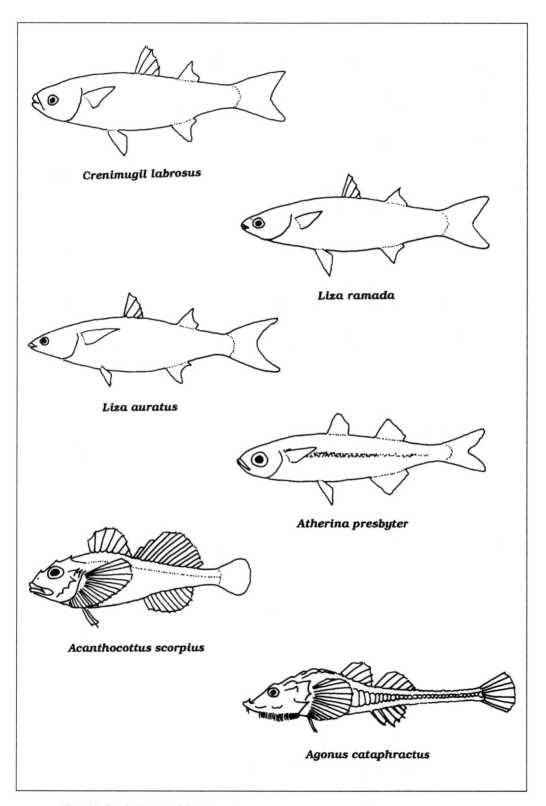

Crenimugil labrosus

Liza ramada

Liza auratus

Atherina presbyter

Acanthocottus scorpius

Agonus cataphractus

Fig. 115. Brackish-water fish (VI).

Gasterosteidae

Pungitius pungitius (10-spined stickleback) (Fig. 116)

This is an essentially freshwater species that occurs, rather uncommonly, in low-salinity lagoons with abundant submerged vegetation over the southern part of the North Sea, and rather more frequently (although still far from commonly) in estuaries and brackish ponds over the northern part. When in brackish water the diet is mainly crustacean. Up to 7 cm long.

Gasterosteus aculeatus (3-spined stickleback) (Fig. 116)

The 3-spined is the commonest of the region's sticklebacks in brackish waters, as indeed in fresh waters. It occurs abundantly in estuaries and in all forms of lagoon, and populations may be resident in all these habitats. In lagoons, it is especially associated with beds of submerged vegetation such as *Ruppia*, seldom straying into open water; therefore it does not overlap much with the similarly sized *Pomatoschistus*. Its diet may largely be small crustaceans, although small gastropods and insect larvae are also taken. Different races have been described on the basis of the development of the lateral bony plates: effectively all the races occur in brackish water. Up to 6 cm long.

Spinachia spinachia (15-spined stickleback) (Fig. 116)

In marked contrast to *Pungitius* above, *Spinachia* is essentially a coastal marine stickleback. It is commonest in the north of the region and occurs somewhat infrequently in estuaries and estuarine lagoons amongst *Zostera* or fucoid algae. Juveniles may occur in brackish waters in some numbers in the autumn, otherwise such *Spinachia* as occur do so as solitary individuals. Up to 20 cm long.

Heterosomata

Bothidae

Scophthalmus maximus (turbot) (Fig. 116) and *Scophthalmus rhombus* (brill) (Fig. 116)

These two prime flatfish may occasionally be found, especially when young, in sandy areas of the mouths of estuaries. Neither is able to penetrate far into brackish water, however, although both occur in the extreme southwest of the Baltic. Adults achieve 80 cm and 60 cm, respectively.

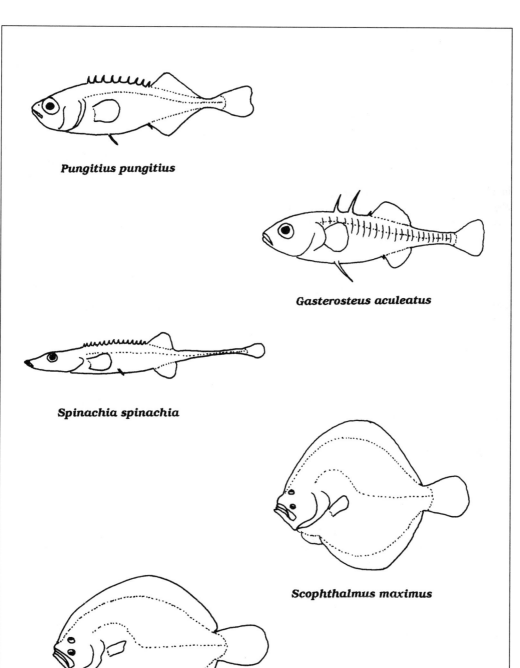

Pungitius pungitius

Gasterosteus aculeatus

Spinachia spinachia

Scophthalmus maximus

Scophthalmus rhombus

Fig. 116. Brackish-water fish (VII).

Pleuronectidae
Pleuronectes platessa (plaice) (Fig. 117)
Plaice sometimes occur in sandy areas in the mouths of estuaries, although their place in brackish water is largely taken by the flounder below. Up to 50 cm long.

Platichthys flesus (flounder) (Fig. 117)
The flounder is the characteristic flatfish of northwest European estuaries and open lagoons, and it can even survive for considerable periods in fresh water. In the Baltic it can breed in (deep) waters of 10‰ salinity. Otherwise, however, populations are not resident in brackish waters; they must leave to breed in the coastal sea. Young flounder of some 10 mm length first enter estuaries and lagoons in the early summer of their first year of life, and there feed on amphipods, polychaetes and small bivalve molluscs, growing to some 80–90 cm length by late autumn when they return to the sea. Thereafter, every year, they re-invade brackish waters in March/April for the duration of the summer. Up to 50 cm long.

Limanda limanda (dab) (Fig. 117)
Dabs, like plaice, may be found in sandy regions of the mouths of estuaries, particularly during the summer months, although, again like the plaice, they are really coastal marine fish. Up to 40 cm long.

Soleidae
Solea solea (Dover sole) (Fig. 117)
The sole is moderately common in the summer months in the more marine sections of estuaries throughout the region except in the extreme north. It is active at night, feeding on amphipods, polychaetes and bivalve molluscs.

Buglossidium luteum (solenette) (Fig. 117)
This small sole is essentially a deep-water species, although it does occur in some shallow, brackish-water regions around Denmark.

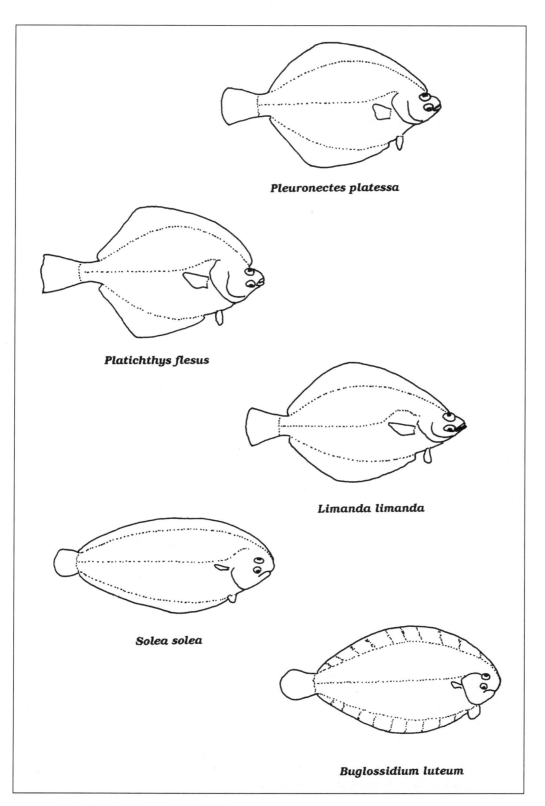

Pleuronectes platessa

Platichthys flesus

Limanda limanda

Solea solea

Buglossidium luteum

Fig. 117. Brackish-water fish (VIII).

References

Ackefors, H. & Hernroth, L. (1972) Djurplankton i östersjöområdet. *Zoologisk Revy,* **34**, 6–31.

Aguesse, P. (1957) La classification des eaux poikilohalines, sa difficulté en Camargue, nouvelle tentative de classification. *Vie et Milieu,* **8**, 341–363.

Amanieu, M. (1967) Introduction à l'étude écologique des réservoirs à poissons de la région d'Arcachon. *Vie et Milieu,* **18**, 381–446.

Ambrose, W.G. (1984) Influences of predatory polychaetes and epibenthic predators on the structure of a soft-bottom community in a Maine estuary. *Journal of Experimental Marine Biology & Ecology,* **81**, 115–145.

André, C. & Rosenberg, R. (1991) Adult–larval interactions in the suspension-feeding bivalves *Cerastoderma edule* and *Mya arenaria. Marine Ecology Progress Series,* **71**, 227–234.

Athersuch, J., Horne, D.J. & Whittaker, J.E. (1989) *Marine and Brackish Water Ostracods. Synopses of the British Fauna (New Series)* No. 43. Brill, Leiden.

Axell, H. & Hosking, E. (1977) *Minsmere. Portrait of a Bird Reserve* . Hutchinson, London.

Bachelet, G. & Yacine-Kassab, M. (1987) Intégration de la phase post-recrutée dans la dynamique des populations du gastéropode intertidal *Hydrobia ulvae* (Pennant). *Journal of Experimental Marine Biology & Ecology,* **111**, 37–60.

Bagenal, T.B. (1973) *Identification of British Fishes.* Hulton, Amersham.

Baird, D. & Milne, H. (1981) Energy flow in the Ythan Estuary, Aberdeenshire, Scotland. *Estuarine, Coastal & Shelf Science,* **13**, 455–472.

Ball, I.R. & Reynoldson, T.B. (1981) *British Planarians.* Cambridge University Press, Cambridge.

Bamber, R.N., Batten, S.D., Sheader, M. & Bridgwater, N.D. (1992) On the ecology of brackish water lagoons in Great Britain. *Aquatic Conservation,* **2**, 65–94.

Bank, R.A. & Butot, L.J.M. (1984) Some more data on *Hydrobia ventrosa* (MONTAGU, 1803) and 'Hydrobia ' *stagnorum* (GMELIN, 1791) with remarks on the genus *Semisalsa* RADOMAN, 1974 (Gastropoda, Prosobranchia, Hydrobioidea). *Malakologische Abhandlungen, Staatliches Museum für Tierkunde Dresden,* **10**, 5–15.

Barnes, R.S.K. (1980) *Coastal Lagoons.* Cambridge University Press, Cambridge.

Barnes, R.S.K. (1984) *Estuarine Biology,* 2nd edn. Arnold, London.

Barnes, R.S.K. (1988) The faunas of land-locked lagoons: chance differences and the problems of dispersal. *Estuarine, Coastal & Shelf Science,* **26**, 309–318.

Barnes, R.S.K. (1989a) What, if anything, is a brackish-water fauna? *Transactions of the Royal Society of Edinburgh, Earth Sciences,* **80**, 235–240.

Barnes, R.S.K. (1989b) The coastal lagoons of Britain: an overview and conservation appraisal. *Biological Conservation,* **49**, 295–313.

Barnes, R.S.K. (1989c) Marine animals of Blakeney Point and Scolt Head Island. In: Allison, H. & Morley, J. (eds.), *Blakeney Point and Scolt Head Island,* 5th edn., pp. 67–75. The National Trust, Norfolk.

Barnes, R.S.K. (1991) Dilemmas in the theory and practice of biological conservation as exemplified by British coastal lagoons. *Biological Conservation,* **55**, 315–328.

Barnes, R.S.K. (1994a) A critical appraisal of the application of Guélorget and Perthuisot's concepts of 'the paralic ecosystem' and 'confinement' to macrotidal Europe. *Estuarine, Coastal and Shelf Science,* **38**, 41–48.

Barnes, R.S.K. (1994b) Macrofaunal community structure and life histories in coastal lagoons. In: Kjerfve, B. (ed.) *Coastal Lagoon Processes.* Elsevier, Amsterdam, in press.

Barnes, R.S.K. & Heath, S.E. (1980) The shingle foreshore / lagoon system of Shingle Street, Suffolk: a preliminary survey. *Transactions of the Suffolk Naturalists' Society,* **18**, 168–181.

Barnes, R.S.K. & Hughes, R.N. (1988) *An Introduction to Marine Ecology,* 2nd edn. Blackwell Scientific, Oxford.

Barnes, R.S.K., Calow, P. & Olive, P.J.W. (1988) *The Invertebrates: A New Synthesis.* Blackwell Scientific, Oxford.

Barrett, J.H. & Yonge, C.M. (1958) *Collins Pocket Guide to the Sea Shore.* Collins, London.

Barth, G.M. & Anthon, H. (1956) *Hvad Finder jeg pa Stranden.* Politikens, Copenhagen.

Bassindale, R. (1964) *British Barnacles.* Linnean Society, London.

Bianchi, C.N. (1981) *Guide per il Riconoscimento delle Specie Animali delle Acque Lagunari e Costiere Italiane. 5. Policheti Serpuloidei.* Consiglio Nazionale delle Ricerche, Italy.

Billheimer, L.E. & Coull, B.C. (1988) Bioturbation and recolonization of meiobenthos in juvenile spot (Pisces) feeding pits. *Estuarine, Coastal and Shelf Science,* **27**, 335–340.

Bratton, J.H. (ed.) (1991) *British Red Data Books. 3. Invertebrates other than Insects.* Joint Nature Conservation Committee, Peterborough.

Brinkhurst, R.O. (1982) *British and Other Marine and Estuarine Oligochaetes. Synopses of the British Fauna (New Series)* No. 21. Cambridge University Press, Cambridge.

Brunberg, L. (1964) On the nemertean fauna of Danish waters. *Ophelia,* **1**, 77–111.

Bulger, A.J., Hayden, B.P., Monaco, M.E., Nelson, D.M. & McCormick-Ray, M.G. (1993) Biologically-based estuarine salinity zones derived from a multivariate analysis. *Estuaries,* **16**, 311–322.

Carter, R.W.G. (1988) *Coastal Environments.* Academic Press, London.

Chevreux, E. & Fage, L. (1925) *Faune de France. 9. Amphipodes.* Librairie de la Faculté des Sciences, Paris.

Christiansen, M.E. (1969) *Crustacea Decapoda Brachyura.* Universitetsforlaget, Oslo.

Clapham, A.R., Tutin, T.G. & Warburg, E.F. (1965) *Flora of the British Isles. Illustrations Part IV. Monocotyledones.* Cambridge University Press, Cambridge.

Clarke, R.O.S. (1973) Coleoptera. Family Heteroceridae. *Royal Entomological Society of London Handbooks for the Identification of British Insects,* **5**(2c), 1–14.

Clegg, J. (1956) *Pond Life.* Warne, London.

Coles, S.M. (1979) Benthic microalgal populations on intertidal sediments and their role as precursors to salt marsh development. In: Jefferies, R.L. & Davy, A.J. (eds), *Ecological Processes in Coastal Environments*, pp. 25–42. Blackwell Scientific, Oxford.

Comin, F.A. & Northcote, T.G. (eds) (1990) *Saline Lakes. Hydrobiologia,* Vol. 197.

Cranston, P.S. (1982) *A Key to the larvae of the British Orthocladiinae (Chironomidae).* Freshwater Biological Association, Ambleside.

Davidson, N.C. *et al.* (1991) *Nature Conservation and Estuaries in Great Britain.* Nature Conservancy Council, Peterborough.

Davies, J.L. (1972) *Geographical Variation in Coastal Development.* Oliver & Boyd, Edinburgh

de Jonge, V.N. (1974) Classification of brackish coastal inland waters. *Hydrobiological Bulletin,* **8**, 29–39.

den Hartog, C. (1964) Typologie des Brackwassers. *Helgolander Wissenschaftliche Meeresuntersuchungen,* **10**, 377–390.

den Hartog, C. (1974) Brackish-water classification, its development and problems. *Hydrobiological Bulletin,* **8**, 15–28.

Dorey, A.E., Little, C. & Barnes, R.S.K. (1973) An ecological study of the Swanpool, Falmouth. II. Hydrography and its relation to animal distributions. *Estuarine and Coastal Marine Science,* **1**, 153–176.

Ehlers, J. (1988) *The Morphodynamics of the Wadden Sea.* Balkema, Rotterdam.

Elmgren, R., Ankar, S., Marteleur, B. & Ejdung, G. (1986) Adult interference with postlarvae in soft sediments: the *Pontoporeia–Macoma* example. *Ecology,* **67**, 827–836.

Escaravage, V. & Castel, J. (1990) The impact of the lagoonal shrimp *Palaemonetes varians* (Leach) on meiofauna in a temperate coastal impoundment. *Acta Oecologica,* **11**, 409–418.

Evans, S. (1983) Production, predation and food niche segregation in a marine shallow soft-bottom community. *Marine Ecology Progress Series,* **10**, 147–157.

Fauchald, K. (1977) The polychaete worms: definitions and keys to the orders, families and genera. *Natural History Museum of Los Angeles County, Science Series,* **28**, 1–190.

Fauvel, P. (1923) *Faune de France. 5. Polychètes Errantes.* Lechevalier, Paris.

Fauvel, P. (1927) *Faune de France. 16. Polychètes Sédentaires.* Lechevalier, Paris.

Fenchel, T. (1975) Factors determining the distribution patterns of mud snails (Hydrobiidae). *Oecologia, Berlin,* **20**, 1–17.

Fenchel, T. & Kolding, S. (1979) Habitat selection and distribution patterns of five species of the amphipod genus *Gammarus. Oikos,* **33**, 316–322.

Fenchel, T. & Riedl, R.J. (1970) The sulphide system: a new biotic community underneath the oxidised layer of marine sand bottoms. *Marine Biology,* **7**, 255–268.

Fish, J.D. & Fish, S. (1977) The veliger larva of *Hydrobia ulvae* with observations on the veliger of *Littorina littorea* (Mollusca: Prosobranchia). *Journal of Zoology, London,* **182**, 495–503.

Fish, J.D. & Fish, S. (1989) *A Student's Guide to the Seashore.* Unwin Hyman, London.

Fisher, J.J. (1980) Shoreline erosion, Rhode Island and North Carolina coasts. In: Schwartz, M.L. & Fisher, J.J. (eds), *Proceedings of the Per Bruun Symposium,* pp. 32–54. University of Rhode Island, Newport.

Fitter, R. & Manuel, R. (1986) *Collins Field Guide to Freshwater Life.* Collins, London.

Flach, E.C. (1992) The influence of four macrozoobenthic species on the abundance of the amphipod *Corophium volutator* on tidal flats of the Wadden Sea. *Netherlands Journal of Sea Research,* **29**, 379–394.

Forsman, B. (1972) Evertebrater vid Svenska Östersjökusten. *Zoologisk Revy,* **34**, 32–56.

Fretter, V. & Graham, A. (1962) *British Prosobranch Molluscs.* The Ray Society, London.

Friday, L.E. (1988) A key to the adults of British water beetles. *Field Studies,* **7**, 1–152.

Gibbs, P.E. (1977) *British Sipunculans.* Academic Press, London.

Gibson, R. (1982) *British Nemerteans.* Cambridge University Press, Cambridge.

Gledhill, T., Sutcliffe, D.W. & Williams, W.D. (1976) *Key to British Freshwater Crustacea: Malacostraca.* Freshwater Biological Association, Windermere.

Goss-Custard, J.D. (1969) The winter feeding ecology of the redshank *Tringa totanus. Ibis,* **111**, 338–356.

Graham, A. (1988) *Molluscs: Prosobranch and Pyramidellid Gastropods.* Brill/Backhuys, Leiden.

Green, J. (1968) *The Biology of Estuarine Animals.* Sidgwick & Jackson, London.

Guélorget, O. & Perthuisot, J-P. (1983) Le domaine paralique. Expressions géologiques et économiques du confinement. *Travaux du Laboratoire de Géologie de l'École Normale Supériere,* **16**, 136 pp.

Guélorget, O. & Perthuisot, J-P. (1992) Paralic ecosystems. Biological organization and functioning. *Vie et Milieu,* **42**, 215–251.

Hammond, C.O. (1983) *The Dragonflies of Great Britain and Ireland,* 2nd Edn. Harley, Colchester.

Hansen, M. (1987) The Hydrophiloidea (Coleoptera) of Fennoscandia and Denmark. *Fauna Entomologica Scandinavica,* **18**, 254 pp.

Harde, K.W. (1981) *Der Kosmos–Käferführer.* Keller, Stuttgart.

Hardisty, J. (1990) *The British Seas.* Routledge, London.

Harris, T. (1980) Invertebrate community structure on, and near, Bull Hill in the Exe Estuary. In: Boalch, G.T. (ed.), *Essays on the Exe Estuary,* pp. 135–158. Devonshire Association for the Advancement of Science, Literature and Art, Exeter.

Hartmann-Schröder, G. (1971) *Die Tierwelt Deutschlands. 58. Annelida, Borstenwürmer, Polychaeta.* Fischer, Jena.

Hayward, P.J. (1985) *Ctenostome Bryozoans.* Brill/Backhuys, Leiden.

Hayward, P.J. & Ryland, J.S. (1979) *British Ascophoran Bryozoans.* Academic Press, London.

Hayward, P.J. & Ryland, J.S. (1990) *The Marine Fauna of the British Isles and North-West Europe* (2 vols). Clarendon, Oxford.

Head, P.C. (1972) Nutrient studies. In: James, A. (ed.), *Pollution of the River Tyne Estuary,* pp. 13–17. University of Newcastle upon Tyne, Department of Civil Engineering, Bulletin No. 42.

Heerebout, G.R. (1970) A classification system for isolated brackish inland waters, based on median chlorinity and chlorinity fluctuation. *Netherlands Journal of Sea Research,* **4**, 494–503.

Hickin, N.E. (1967) *Caddis Larvae. Larvae of the British Trichoptera.* Hutchinson, London.

Hiscock, S. (1979) A field key to the British brown seaweeds (Phaeophyta). *Field Studies,* **5**, 1–44.

Holdich, D.M. & Jones, J.A. (1983) *Tanaids.* Cambridge University Press, Cambridge.

Holmen, M. (1987) The aquatic Adephaga (Coleoptera) of Fennoscandia and Denmark. I. Gyrinidae, Haliplidae, Hygrobiidae and Noteridae. *Fauna Entomologica Scandinavica,* **20**, 168 pp.

Hunt, O.D. (1971) Holkham Salts Hole, an isolated salt-water pond with relict features. An account based on studies by the late C.F.A. Pantin. *Journal of the Marine Biological Association of the U.K.,* **51**, 717–741.

Ingle, R.W. (1983) *Shallow-water Crabs.* Cambridge University Press, Cambridge.

IUCN (1983) *The IUCN Invertebrate Red Data Book.* IUCN, Gland, Switzerland.

Jansson, A. (1986) The Corixidae (Heteroptera) of Europe and some adjacent regions. *Acta Entomologica Fennica,* **47**, 1–94.

Janus, H. (1965) *The Young Specialist Looks at Land and Freshwater Molluscs.* Burke, London.

Jensen, K.T. (1985) The presence of the bivalve *Cerastoderma edule* affects migration, survival and reproduction of the amphipod *Corophium volutator. Marine Ecology Progress Series,* **25**, 269–277.

Jensen, K.T. & Jensen, J.N. (1985) The importance of some epibenthic predators on the density of juvenile benthic macrofauna in the Danish Wadden Sea. *Journal of Experimental Marine Biology & Ecology,* **89**, 157–174.

Jensen, K.T. & Kristensen, L.D. (1990) A field experiment on competition between *Corophium volutator* (Pallas) and *Corophium arenarium* Crawford (Crustacea: Amphipoda): effects on survival, reproduction and recruitment. *Journal of Experimental Marine Biology & Ecology,* **137**, 1–24.

Jones, A.M. & Baxter, J.M. (1987) *Molluscs: Caudofoveata, Solenogastres, Polyplacophora and Scaphopoda.* Brill/Backhuys, Leiden.

Jones, N.S. (1976) *British Cumaceans.* Academic Press, London.

Joy, N.H. (1932) *A Practical Handbook of British Beetles.* Witherby, London.

Klausnitzer, B. (1978) *Ordnung Coleoptera (Larven).* Junk, The Hague.

Klausnitzer, B. (1991) *Die Larven der Käfer Mitteleuropas. I. Adephaga.* Goecke & Evers, Krefeld.

Kneib, R.T. (1988) Testing for indirect effects of predation in an intertidal soft-bottom community. *Ecology,* **69**, 1795–1805.

Kolding, S. (1986) Interspecific competition for mates and habitat selection in five species of *Gammarus* (Amphipoda: Crustacea). *Marine Biology,* **91**, 491–495.

Kristensen, E. (1988) Factors influencing the distribution of nereid polychaetes in Danish coastal waters. *Ophelia,* **29**, 127–140.

Kühl, H. (1972) Hydrography and biology of the Elbe Estuary. *Oceanography and Marine Biology. An Annual Review,* **10**, 225–309.

Lee, A.J. & Ramster, J.W. (eds) (1981) *Atlas of the Seas Around the British Isles.* MAFF, Lowestoft.

Levinton, J.S. (1972) Stability and trophic structure in deposit-feeding and suspension-feeding communities. *American Naturalist,* **106**, 472–486.

Lincoln, R.J. (1979) *British Marine Amphipoda: Gammaridea.* British Museum (Natural History), London.

Lincoln, R.J., Boxshall, G.A. & Clark, P.F. (1982) *A Dictionary of Ecology, Evolution and Systematics.* Cambridge University Press, Cambridge.

Little, C., Barnes, R.S.K. & Dorey, A.E. (1973) An ecological study of the Swanpool, Falmouth. 3. Origin and history. *Cornish Studies,* **1**, 33–48.

Little, C., Seaward, D.R. & Williams, G.A. (1989) Distribution of intertidal molluscs in lagoonal shingle (The Fleet, Dorset, U.K.). *Journal of Conchology,* **33**, 225–232.

Macan, T.T. (1959) *A Guide to Freshwater Invertebrate Animals.* Longmans, London.

McIntosh, W.C. (1873–1923) *A Monograph of the British Annelids.* Ray Society, London.

Maciolek, N.J. (1984) New records and species of *Marenzelleria* Mesnil and *Scolecolepides* Ehlers (Polychaeta; Spionidae) from northeastern North America. In: Hutchings, P.A. (ed.), *Proceedings of the First International Polychaete Conference, Sydney,* pp. 48–62. Linnean Society of New South Wales, Sydney.

Makings, P. (1977) A guide to the British coastal Mysidacea. *Field Studies,* **4**, 575–596.

Mangelsdorf, P.C. (1967) Salinity measurement in estuaries. In: Lauff, G.H. (ed.), *Estuaries,* pp. 71–79. American Association for the Advancement of Science, Washington D.C.

Manuel, R.L. (1988) *British Anthozoa,* revised Edn. Brill/Backhuys, Leiden.

Merritt, R.W. & Cummins, K.W. (1984) *An Introduction to the Aquatic Insects of North America,* 2nd edn. Kendall/Hunt, Iowa.

Millar, R.H. (1970) *British Ascidians.* Academic Press, London.

Millard, N.A.H. (1975) Monograph of the Hydroida of southern Africa. *Annals of the South African Museum,* **68**, 1–513.

Moore, J.A. (1986) *Charophytes of Great Britain and Ireland.* Botanical Society of the British Isles, London.

Morri, C. (1981) *Guide per il Riconoscimento delle Specie Animali delle Acque Lagunari e Costiere Italiane.* **6**. *Idrozoi Lagunari.* Consiglio Nazionale delle Ricerche, Italy.

Morrisey, D.J. (1988) Differences in effects of grazing by deposit-feeders *Hydrobia ulvae* (Pennant) (Gastropoda: Prosobranchia) and *Corophium arenarium* Crawford (Amphipoda) on sediment microalgal populations. I. Qualitative differences. *Journal of Experimental Marine Biology & Ecology,* **118**, 33–42.

Mortensen, T. (1927) *Handbook of the Echinoderms of the British Isles.* Oxford University Press, London.

Mundy, S.P. (1980) *A Key to the British and European Freshwater Bryozoans.* Freshwater Biological Association, Windermere.

Munksgaard, C. (1990) Electrophoretic separation of morphologically similar species of the genus *Rissoa* (Gastropoda: Prosobranchia). *Ophelia,* **31**, 97–104.

Muus, B.J. (1963) Some Danish Hydrobiidae with the description of a new species, *Hydrobia neglecta. Proceedings of the Malacological Society of London,* **35**, 131–138.

Muus, B.J. (1967) The faunas of Danish estuaries and lagoons. *Meddelelser fra Danmarks Fiskeri- og Havundersøgelser (Ny Serie),* **5**, 1–316.

Naylor, E. (1972) *British Marine Isopods.* Academic Press, London.

Nelson, W.G., Cairns, K.D. & Virnstein, R.W. (1982) Seasonality and spatial patterns of seagrass-associated amphipods of the Indian River Lagoon, Florida. *Bulletin of Marine Science*, **32**, 121–129.

Nelson-Smith, A. (1965) Marine biology of Milford Haven: the physical environment. *Field Studies*, **2**, 155–188.

Nyst, P-H. (1835) (Séance du 4 Juillet; Lecture) Mollusques. *Bulletins de l'Académie Royale des Sciences et Belles-Lettres de Bruxelles*, **2**, 235–236.

Occhipinti Ambrogi, A. (1981) *Guide per il Riconoscimento delle Specie Animali delle Acque Lagunari e Costiere Italiane. 7. Briozoi Lagunari*. Consiglio Nazionale delle Ricerche, Italy.

Olafsson, E.B. (1986) Density dependence in suspension-feeding and deposit-feeding populations of the bivalve *Macoma balthica*: a field experiment. *Journal of Animal Ecology*, **55**, 517–526.

Olafsson, E.B. (1989) Contrasting influences of suspension-feeding and deposit-feeding populations of *Macoma balthica* on infaunal recruitment. *Marine Ecology Progress Series*, **55**, 171–179.

Olafsson, E.B. & Persson, L.E. (1986) The interaction between *Nereis diversicolor* and *Corophium volutator* Pallas as a structuring force in a shallow brackish sediment. *Journal of Experimental Marine Biology & Ecology*, **103**, 103–117.

Palmer, J. (1993) Investigation of the control of mosquitoes by GB-1111 and *Hydrometra stagnorum*. Unpublished University of Cambridge Undergraduate Project Report.

Pashley, H.E. (1985) Feeding and Optimization – The Foraging Behaviour of *Nereis diversicolor* (Polychaeta). Unpublished PhD thesis.

Petersen, G.H. (1958) Notes on the growth and biology of the different *Cardium* species in Danish brackish water areas. *Meddelelser fra Danmarks Fiskeri- og Havundersøgelser (Ny Serie)*, **2**, 3–31.

Petersen, G.H. & Russell, P.J.C. (1971a) *Cardium hauniense* nov. sp. A new brackish water bivalve from the Baltic. *Ophelia*, **9**, 11–13.

Petersen, G.H. & Russell, P.J.C. (1971b) *Cardium hauniense* compared with *C. exiguum* and *C. glaucum*. *Proceedings of the Malacological Society of London*, **39**, 409–420.

Piersma, T. (1987) Production by intertidal benthic animals and limits to their predation by shorebirds: a heuristic model. *Marine Ecology Progress Series*, **38**, 187–196.

Pinder, L.C.V. (1978) *A Key to Adult Males of British Chironomidae* (2 vols). Freshwater Biological Association, Ambleside.

Platt, H.M. & Warwick, R.M. (1983) *Free-living Marine Nematodes. Part I. British Enoplids*. Cambridge University Press, Cambridge.

Platt, H.M. & Warwick, R.M. (1988) *Free-living Marine Nematodes. Part II. British Chromadorids*. Brill/Backhuys, Leiden.

Pleijel, F. & Dales, R.P. (1991) *Polychaetes: British Phyllodocoideans, Typhloscolecoideans and Tomopteroideans*. Universal Book Services, Leiden.

Pollard, M. (1978) *North Sea Surge*. Dalton, Lavenham.

Posey, M.H. & Hines, A.H. (1991) Complex predator–prey interactions within an estuarine benthic community. *Ecology*, **72**, 2155–2169.

Prudhoe, S. (1982) *British Polyclad Turbellarians*. Cambridge University Press, Cambridge.

Quigley, M. (1977) *Invertebrates of Streams and Rivers*. Arnold, London.

Raffaelli, D. & Milne, H. (1987) An experimental investigation of the effects of shorebird and flatfish predation on estuarine invertebrates. *Estuarine, Coastal & Shelf Science*, **24**, 1–13.

Rasmussen, E. (1944) Faunistic and biological notes on marine invertebrates. I. *Videnskabelige Meddelelser fra Dansk Naturhistorisk Forening i Kjobenhavn*, **107**, 207–233.

Rasmussen, E. (1973) Systematics and ecology of the Isefjord marine fauna (Denmark). *Ophelia*, **11**, 1–507.

Reay, P.J. & Cornell, V. (1988) Identification of grey mullet (Teleostei: Mugilidae) juveniles from British waters. *Journal of Fish Biology*, **32**, 95–99.

Redeke, H.C. (1922) Zur Biologie der niederländischen Brackwassertypen (Ein Beitrag zur regionalen Limnologie). *Bijdragen tot de Dierkunde,* **22**, 329–335.

Rehfeldt, N. (1968) Reproductive and morphological variations in the prosobranch '*Rissoa membranacea*'. *Ophelia,* **5**, 157–173.

Reid, G.K. & Wood, R.D. (1976) *Ecology of Inland Waters and Estuaries,* 2nd edn. Van Nostrand, New York.

Reise, K. (1978) Experiments on epibenthic predation in the Wadden Sea. *Helgolander Wissenschaftliche Meeresuntersuchungen,* **31**, 55–101.

Reise, K. (1985) *Tidal Flat Ecology* . Springer, Berlin.

Reitter, E. (1908) *Fauna Germanica. Die Käfer des Deutschen Reiches.* Lutz, Stuttgart.

Remane, A. & Schlieper, C. (1971) *Biology of Brackish Water,* 2nd edn. (*Die Binnengewässer,* **25**.) Schweizerbart'sche, Stuttgart / Wiley, New York.

Rhoads, D.C. (1974) Organism–sediment relations on the muddy sea floor. *Oceanography and Marine Biology Annual Review,* **12**, 263–300.

Rhoads, D.C. & Young, D.K. (1970) The influence of deposit-feeding organisms on sediment stability and community trophic structure. *Journal of Marine Research,* **28**, 150–178.

Richardson, H. (1905) A monograph on the isopods of North America. *Bulletin of the United States National Museum,* No. 54, 1–727.

Richoux, P. (1982) Introduction pratique à la systématique des organismes des eaux continentales françaises. 2. Coléoptères aquatiques (genres: adultes et larves). *Bulletin Mensuel de la Société Linnéenne de Lyon,* **51**, 105–128, 257–272 and 289–303.

Rönn, C., Bonsdorff, E. & Nelson, W.G. (1988) Predation as a mechanism of interference within infauna in shallow brackish water soft bottoms: experiments with an infaunal predator, *Nereis diversicolor* O.F. Müller. *Journal of Experimental Marine Biology & Ecology,* **116**, 143–157.

Russell, F.S. (1953) *The Medusae of the British Isles. Anthomedusae, Leptomedusae, Limnomedusae, Trachymedusae and Narcomedusae.* Cambridge University Press, Cambridge.

Ryland, J.S. & Hayward, P.J. (1977) *British Anascan Bryozoans.* Academic Press, London.

Sars, G.O. (1895) *An Account of the Crustacea of Norway. 1. Amphipoda.* Cammermeyers, Christiania.

Savage, A.A. (1989) *Adults of the British Aquatic Hemiptera Heteroptera.* Freshwater Biological Association, Ambleside.

Schäfer, W. (1972) *Ecology and Palaeoecology of Marine Environments* (transl. Oertel, I.). Oliver & Boyd, Edinburgh.

Seaward, D.R. (1988) *Caecum armoricum* de Folin 1869, new to the British marine molluscan fauna, living in the Fleet, Dorset, within an unusual habitat. *Proceedings of the Dorset Natural History and Archaeological Society,* **109**, 165.

Shirt, D.B. (1987) *British Red Data Books. 2. Insects.* Nature Conservancy Council, Peterborough.

Smaldon, G. (1979) *British Coastal Shrimps and Prawns.* Academic Press, London.

Smidt, E.L.B. (1951) Animal production in the Danish Waddensea. *Meddelelser fra Kommissionen for Danmarks Fiskeri og Havundersøgelser (Serie Fiskeri),* **11**, 1–151.

Smith, K.G.V. (1989) *An Introduction to the Immature Stages of British Flies. Handbooks for the Identification of British Insects,* **10**(14). Royal Entomological Society of London.

Tattersall, W.M. & Tattersall, O.S. (1951) *The British Mysidacea.* The Ray Society, London.

Tebble, N. (1966) *British Bivalve Seashells.* British Museum (Natural History), London.

Thistle, D. (1981) Natural physical disturbance and communities in marine soft bottoms. *Marine Ecology Progress Series,* **6**, 223–228.

Thompson, T.E. (1988) *Molluscs: Benthic Opisthobranchs.* Brill/Backhuys, Leiden.

Thorson, G. (1946) Reproduction and larval development of Danish marine bottom invertebrates, with special reference to the planktonic larvae in the Sound (Øresund). *Meddelelser fra Kommissionen for Danmarks Fiskeri og Havundersøgelser, Serie Plankton,* **4**, 1–523.

Ushakov, P.V. (1955) *Mnogoshchetinkovye Chervi Dal'nevostochnykh Morei SSSR*. Izdatel'stvo Akademii Nauk SSSR, Moscow.

van Urk, R.M. (1987) *Ensis americanus* (Binney) (Syn. *E. directus* auct. non Conrad). A recent introduction from Atlantic North-America. *Journal of Conchology, 32*, 329–333.

Verdonschot, P.F.M., Smies, M. & Sepers, A.B.J. (1982) The distribution of aquatic oligochaetes in brackish inland waters in the SW Netherlands. *Hydrobiologia, 89*, 29–38.

Verhoeven, J.T.A. (1980) The ecology of *Ruppia*-dominated communities in western Europe. II. Synecological classification. Structure and dynamics of the macroflora and macrofauna communities. *Aquatic Botany, 8*, 1–85.

Vlas, J. de (1979) Annual food intake by plaice and flounder in a tidal flat area in the Dutch Wadden Sea, with special reference to consumption of regenerating parts of macrobenthic prey. *Netherlands Journal of Sea Research, 13*, 117–153.

Wagret, P. (1959) *Les Polders*. Dunod, Paris.

Wallace, I.D., Wallace, B. & Philipson, G.N. (1990) *A Key to the Case-bearing Caddis Larvae of Britain and Ireland*. Freshwater Biological Association, Ambleside.

Westheide, W. (1990) *Polychaetes: Interstitial Families*. Universal Book Services, Leiden.

Wheeler, A. (1969) *The Fishes of the British Isles and North-West Europe*. Macmillan, London.

Wheeler, A. (1978) *Key to the Fishes of Northern Europe*. Warne, London.

Whittow, J. (1984) *The Penguin Dictionary of Physical Geography*. Penguin Books, Harmondsworth.

Williams, W.D. (ed.) (1981) *Salt Lakes. Hydrobiologia*, Vol. 81/82.

Wilson, E.O. (1992) *The Diversity of Life*. Belknap, Harvard.

Wilson, W.H. (1981) Sediment-mediated interactions in a densely populated infaunal assemblage: the effects of the polychaete *Abarenicola pacifica*. *Journal of Marine Research, 39*, 735–748.

Wilson, W.H. (1989) Predation and the mediation of intraspecific competition in an infaunal community in the Bay of Fundy. *Journal of Experimental Marine Biology & Ecology, 132*, 221–245.

Wiltse, W.I., Foreman, K.H., Teal, J.H. & Valiela, I. (1984) Effects of predators and food resources on the macrobenthos of salt marsh creeks. *Journal of Marine Research, 42*, 923–942.

Witte, F. & Wilde, P.A.W.J. de (1979) On the ecological relation between *Nereis diversicolor* and juvenile *Arenicola marina*. *Netherlands Journal of Sea Research, 13*, 394–405.

Wolff, W.J. (1972) Origin and history of the brackish water fauna of N.W. Europe. In: *Fifth European Marine Biology Symposium*, pp. 11–18. Piccin, Padova.

Wolff, W.J. (1973) The estuary as a habitat. An analysis of data on the soft-bottom macrofauna of the estuarine area of the rivers Rhine, Meuse and Scheldt. *Zoologische Verhandelingen, 126*, 3–242.

Wolff, W.J. (ed.) (1983) *Ecology of the Wadden Sea* (3 vols). Balkema, Rotterdam.

Wolff, W.J. & Wolf, L. de (1977) Biomass and production of zoobenthos in the Grevelingen Estuary, The Netherlands. *Estuarine and Coastal Marine Science, 5*, 1–24.

Woodin, S.A. (1976) Adult–larval interactions in dense infaunal assemblages: patterns of abundance. *Journal of Marine Research, 34*, 25–41.

Woodin, S.A. (1981) Disturbance and community structure in a shallow water sand flat. *Ecology, 62*, 1052–1066.

Woodin, S.A. & Jackson, J.B.C. (1979) Interphyletic competition among marine benthos. *American Zoologist, 19*, 1029–1043.

Yonge, C.M. & Thompson, T.E. (1976) *Living Marine Molluscs*. Collins, London.

Zenkevitch, L. (1963) *Biology of the Seas of the U.S.S.R.* (transl. Botcharskaya, S.). Allen & Unwin, London.

Zwarts, L. & Blomert, A-M. (1992) Why knot *Calidris canutus* take medium-sized *Macoma balthica* when six prey species are available. *Marine Ecology, Progress Series, 83*, 113–128.

Zwarts, L. & Esselink, P. (1989) Versatility of male curlews *Numenius arquata* preying upon *Nereis diversicolor*: deploying contrasting capture modes dependent on prey availability. *Marine Ecology*

Progress Series, **56**, 255–269.

Zwarts, L. & Wanink, J. (1989) Siphon size and burying depth in deposit- and suspension-feeding benthic bivalves. *Marine Biology,* **100**, 227–240.

Zwarts, L., Blomert, A-M. & Wanink, J.H. (1992) Annual and seasonal variation in the food supply harvestable by knot *Calidris canutus* staging in the Wadden Sea in late summer. *Marine Ecology Progress Series,* **83**, 129–139.

Index to organisms

(Citations to illustrations are in **bold**)

Abra, 34
Abra alba, 143, 150, **151**
Abra tenuis, 143, 150, **151**
Acanthocottus scorpius, 257, 266, **267**
Acanthodoris pilosa, 113, 134, **135**
Acentropus niveus, 208, 224, **225**
Acmaeidae, 115
Actinia equina, 59, 66, **67**
actiniarians, 66–9
Actiniidae, 66
Adalaria proxima, 113, 134, **135**
Aegires punctilucens, 113, 136, **137**
Aeolidiella, 112, 136
Aeolidiella alderi, 136
Aeolidiella glauca, 136
Aeolidiella sanguinea, 136
Aeolidiidae, 136
Agabus, 209, **233**
Agabus biguttatus, 215, 232, **233**
Agabus bipustulatus, 215, 232, **233**
Agabus conspersus, 215, 232, **233**
Agabus nebulosus, 215, 232, **233**
Agnatha, 258
Agonidae, 266
Agonus cataphractus, 257, 266, **267**
Akera bullata, 105, 130, **131**
Akera bullata var. *farrani*, 130
Akera bullata var. *nana*, 130
Akeridae, 130
Aktedrilus, 104
Alcyonidiidae, 246
Alcyonidium diaphanum, 246
Alcyonidium gelatinosum, 245, 246, **247**
Alcyonidium polyoum, see Alcyonidium gelatinosum
Alderia modesta, 112, 132, **133**
algae, 16, 24, 31, 32, 34, 39, 46, 62, 64, 68, 94, 126, 130, 132, 148, 154, 166, 168, 169, 180, 184, 185, 188, 192, 194, 198, 200, 220, 230, 242, 246, 248, 252, 266, 268
Alkmaria romijni, 31, 84, 98, **99**
allis shad, *see Alosa alosa*
allogastropod molluscs, 128
Alosa alosa, 257, 258, **259**
Alosa fallax, 257, 258, **259**
Amathia lendigera, 245, 246, **247**
American jack-knife clam, *see Ensis directus*
Ampelisca brevicornis, 172, 178, **179**

Ampeliscidae, 178
Ampharete acutifrons, see Ampharete grubei
Ampharete grubei, 84, 98, **99**
Ampharetidae, 78, 98
Amphichaeta sannio, 104
Amphilochidae, 178
Amphipholis squamata, 250, **251**
amphipod crustaceans, 25, 35, 43, 45, 52, 170–200, 260, 262, 264, 270
Amphiporidae, 74
Amphiporus lactifloreus, 70, 74, **75**
Amphitrite figulus, 84, 86, 100, **101**, 128
Amphiuridae, 250
Ampithoe rubricata, 174, 192, **193**
Ampithoidae, 192
Anacanthini, 262
Anaitides groenlandica, see Phyllodoce groenlandica
Anaitides maculata, see Phyllodoce maculata
Anaspida, 130
anchovy, *see Engraulis encrasicolus*
Anguilla anguilla, 254, 260, **261**
Anguillidae, 260
Anisobranchida, 116
annelids, 78–104
Anomiidae, 146
anthozoans, 66–9
Anthuridae, 164
Anurida maritima, 207, 218, **219**
Aora gracilis, see Aora typica
Aora typica, 174, 194, **195**
Aoridae, 194
Apherusa, 173, 192, **193**
Aphrodita aculeata, 78, 86, **87**
Aphroditidae, 86
Apodes, 260
Apseudes latreillii, 168, **179**
Apseudidae, 168
Archaeopulmonata, 138
Archidorididae, 136
Archidoris pseudoargus, 113, 136, **137**
Arenicola marina, 5, 21, **22**, 23, **24**, 30, 35, 38, 43–5, 80, 86, 96, **97**, 264
Arenicolidae, 96
Armandia cirrhosa, 31, 51, 81, 98, **99**
ascidians, 53, 54, 134, 252
Ascophyllum nodosum, 108, 118, 248
Asellidae, 166
Asellus aquaticus, 31, 162, 166, **167**

Assiminea grayana, 110, 124, **125**
Assimineidae, 124
Astarte borealis, 142, 146, **147**
Astartidae, 146
athecate hydroids, 60–3
Atherina presbyter, 256, 266, **267**
Atherinidae, 266
Atyidae, 130
Atylidae, 192
Atylus swammerdami, 170, 192, **193**
Audouinia tentaculata, see Cirriformia tentaculata
Aurelia aurita, 56, 59, 66, **67**
Aureliidae, 66
avocet, 35

bacteria, 20, 34, 35
Baetidae, 218
Balanidae, 154
Balanus amphitrite, 154, **155**
Balanus balanoides, see Semibalanus balanoides
Balanus improvisus, 154, **155**
barnacles, 22, 54, 154–5
Barnea candida, 142, 146, 152, **153**
Basommatophora, 138
bass, *see Dicentrarchus labrax*
Bathyporeia pilosa, 172, 188, **189**
beadlet anemone, *see Actinia equina*
beetles, 25, 31, 35, 50, 228–43
Belone bellone, 254, 260, **261**
Belonidae, 260
Bembidion laterale, see Cillenus laterale
Berosus, 210, **237**
Berosus affinis, 214, 236, **237**
Berosus spinosus, 51, 214, 236, **237**
Berthella plumula, 114, 130, **131**
bib, *see Trisopterus luscus*
Bimeria franciscana, 58, 62, **63**
birds, 14, 31, 35, 36, 38, 40, 41
Bittium reticulatum, 108, 126, **127**
bivalve molluscs, 23, 34, 35, 38, 42, 43, 54, 138–53, 264, 270
black goby, *see Gobius niger*
black-tailed skimmer, *see Orthetrum cancellatum*
Bledius, 209, 212, 240
Bledius spectabilis, 212, 240, **241**
bloodworm, *see Glycera capitata*
blue-rayed limpet, *see Helcion pellucidum*

blue-tail damselfly, *see Ischnura elegans*
blunt-gaper, *see Mya truncata*
Bothidae, 268
Botryllus schlosseri, 130, 134, 252, **253**
Bougainvilliidae, 62
Bowerbankia caudata, see Bowerbankia gracilis
Bowerbankia gracilis, 246, 248, **249**
Bowerbankia imbricata, 246, 248, **249**
Brachystomia eulimoides, 110, 128, **129**
Brachystomia rissoides, 45, 110, 128, **129**
brill, *see Scophthalmus rhombus*
brittlestars, 250
broad-nosed pipefish, *see Siphonostoma typhle*
brown china mark moth, *see Nymphula nympheata*
brown shrimp, *see Crangon crangon*
bryozoans, 22, 25, 34, 53, 54, 134, 244–50
Buccinidae, 126
Buccinum undatum, 107, 126, **127**
Buglossidium luteum, 255, 270, **271**
Bugula, 244, 250, **251**
Bugulidae, 250
bullhead, *see Agonus cataphractus*

caddis-flies, *see* Trichoptera
Cadlina laevis, 113, 136, **137**
Cadlinidae, 136
Caecidae, 126
Caecum armoricum, 22, 51, 107, 126, **127**
Callicorixa praeusta, 217, 220, **221**
Calliopiidae, 192
Calliopius laeviusculus, 173, 192, **193**
Callopora aurita, 244, 250, **251**
Calloporidae, 250
Campanulariidae, 64
Capitella capitata, 40, 80, 96, **97**
Capitellida, 96
Capitellidae, 96
Carabidae, 209, 212, 242
Carcinides maenas, see Carcinus maenas
Carcinus maenas, 45, 200, 204, **205**, 262
Cardiidae, 148
Cardium, see Cerastoderma
Cardium lamarcki, see Cerastoderma glaucum
caterpillars, *see* Lepidoptera
catworm, *see Nephtys*
Caulleriella zetlandica, 81, 96, **97**
Celleporidae, 250
Cephalaspida, 128
Cephalothricidae, 72
Cephalothrix linearis, 71, 72, **73**
Cephalothrix rufifrons, 71, 72, **73**
Cerastoderma, 33, 34, 36, 39, 40, 42–4, 264
Cerastoderma edule, 31, 43, 45, 46, 141, 148, **149**
Cerastoderma glaucum, 25, 27, 31, 32, 45, 46, 140, 141, 148, **149**
Ceratia proxima, 110, 122, **123**

Cereus pedunculatus, 59, 68, **69**, 136
Cerithiidae, 126
Chaetogammarus marinus, 175, 184, **187**
Chaetogammarus pirloti, 175, 185, **187**
Chaetogammarus stoerensis, 174, 185, **187**
Chaetogaster, 104
Chaetomorpha, 25, 40, 62, 68, 100, 132, 146, 148, 166, 220, 226, 260
charophytes, 25, 51, 100, 148
Cheilostomata, 248–50
Cheirocratus sundevalli, 173, 188, **189**
Chelura terebrans, 170, 198, **199**
Cheluridae, 198
Chironomidae, 50, 68, 206, **225**, 226, 264
chitons, 52, 54, 105
Chrysomelidae, 240
Cillenus laterale, 212, 242, **243**
Cingula trifasciata, 110, 122, **123**
Cirolanidae, 164
Cirratulidae, 96
Cirratulus cirratus, 82, 96, **97**, 128
Cirriformia tentaculata, 82, 96, **97**, 128
clams, *see* bivalve molluscs
Clava multicornis, 56, 62, **63**
Clavidae, 62
Clavopsella navis, 31, 56, 62, **63**
Clavopsella quadranularia, see Clavopsella navis
Clitellio, 104
Cloeon dipterum, 208, 218, **219**
Clupea harengus, 36, 257, 258, **259**
Clupeidae, 258
cnidarians, 54, 56–69
coat-of-mail shells, *see* chitons
cockles, *see Cerastoderma* spp.
Coelambus, 210, **234**
Coelambus impressopunctatus, 216, 232, **234**
Coelambus parallelogrammus, 216, 232, **234**
Coenagrionidae, 218
Coleoptera, 228–43
Collembola, 218
Collisella tessulata, 107, 114, **115**
Colymbetes fuscus, 209, 215, **231**, 232
common goby, *see Pomatoschistus* spp.
common limpet, *see Patella vulgata*
common shrimp, *see Crangon crangon*
Congeria cochleata, 141, 152, **153**
Conopeum reticulum, 244, 245, 248, **249**
Conopeum seurati, 27, 245, 248, **249**
Corambidae, 134
Corbula gibba, 142, 152, **153**
Corbulidae, 152
Cordylophora caspia, 35, 56, 62, **63**, 136, 194
Coregonidae, 260
Coregonus oxyrinchus, 256, 260, **261**
Corixa affinis, 217, 220, **223**
Corixa panzeri, 217, 220, **223**
Corixidae, 31, 50, 208, 217, 220, **221**, 222

Corophiidae, 194, 196–98
Corophium, 34–6, 38, 40, 41, 44, 76, 170, 176–7, **196**, 226, 242
Corophium acherusicum, 177, **196**, 198
Corophium acutum, 176, **197**, 198
Corophium arenarium, 30, 176, 194, **196**
Corophium bonnellii, 177, **196**, 198
Corophium crassicorne, 177, **197**, 198
Corophium curvispinum, 176, **197**, 198
Corophium insidiosum, 177, 194, **196**
Corophium lacustre, 51, 176, **197**, 198
Corophium multisetosum, 176, **197**, 198
Corophium volutator, 30, 176, 194, **196**
Corynidae, 60
Cottidae, 266
crabs, 39, 200–5, 260
Crangon crangon, 39, 200, 204, **205**, 262, 264
Crangonidae, 204
Crassostrea gigas, 140, 144, **145**
Crenimugil labrosus, 256, 266, **267**
Cryptosula pallasiana, 245, 250, **251**
Cryptosulidae, 250
Ctenostomata, 246, 248
cumacean crustaceans, 52, 162
curlew, 35
Cyathura carinata, 162, 164, **165**
Cyclostomata, 258

dab, *see Limanda limanda*
dahlia anemone, *see Urticina felina*
decapod crustaceans, 53, 200–5, 260
Dexamine spinosa, 170. 192, **193**
Dexaminidae, 192
Diadumene cincta, 59, 68, **69**, 136
Diadumene luciae, see Haliplanella lineata
Diadumenidae, 66
Diaphana minuta, 106, 128, **129**
Diaphanidae, 128
Diastylidae, 162
Diastylis rathkei, 162, **165**
diatoms, 20, 34, 35, 44
Dicentrarchus labrax, 36, 257, 264, **265**
Diptera, 30, 50, 226
diving beetles, *see* Dytiscidae
Docoglossida, 115
dogwhelk, *see Nucella lapillus*
Dolichopodidae, 50, 207, 226, **227**, 242
Doridella batava, 35, 114, 134, **135**
Dotidae, 134
Doto coronata, 113, 134, **135**
Dover sole, *see Solea solea*
dragonflies, *see* Odonata
Dreissenidae, 152
Dugesia lugubris, 70, 72
Dugesia polychroa, 70, 72, **73**
Dugesiidae, 72
Dynamena pumila, 58, 64, **65**, 134
Dytiscidae, 232
Dytiscus circumflexus, 209, 215, 232, **234**

echinoderms, 250
edible cockle, *see Cerastoderma edule*
edible winkle, *see Littorina littorea*
Edwardsia ivelli, 31, 51, 58, 68, **69**
Edwardsiidae, 68
eel, *see Anguilla anguilla*
eel-grass, *see Zostera*
eelpout, *see Zoarces viviparus*
Electra crustulenta, 244, 248, **249**
Electra monostachys, 244, 248, **249**
Electra pilosa, 134, 244, 248, **249**
Electridae, 248
Ellobiidae, 138
Elminius modestus, 154, **155**
Elysia viridis, 114, 132, **133**
Elysiidae, 132
Emarginula conica, 107, 114, **115**
Embletonia pallida, *see Tenellia adspersa*
Enchytraeidae, 49, 102, **103**, 104
Engraulis encrasicolus, 256, 258, **259**
Enochrus, 210, **238**
Enochrus bicolor, 214, 236, **238**
Enochrus halophilus, 214, 236, **238**
Enochrus melanocephalus, 214, 236, **238**
Enoplus brevis, 76
Ensis directus, 142, 150, **151**
Enteromorpha, 24, 33, 126, 132, 148
Ephemeroptera, 218
Ephydra riparia, 226
Ephydridae, 50, 206, 226, **227**
Erichthonius, 174, 198, **199**
Erichthonius brasiliensis, *see Erichthonius*
Eriocheir sinensis, 45, 200, 204, **205**
Eteone, 40
Eteone longa, **46**, 83, 88, **89**
Eteone picta, 83, 88, **89**
Eubranchidae, 136
Eubranchus exiguus, 112, 136, **137**
Eubranchus farrani, 112, 136, **137**
Eulalia viridis, 84, 88, **89**
Eulimnogammarus obtusatus, 174, 185, **187**
Eurydice pulchra, 163, 164, **165**
Exogone gemmifera, *see Exogone naidina*
Exogone naidina, 83, 88, **89**

Fabricia sabella, *see Fabricia stellata*
Fabricia stellata, 84, 100, **101**
Fabriciola baltica, 84, 100, **101**
Facelina, 112
false caddis-fly moth, *see Acentropus niveus*
fanworm, *see Sabella pavonina*
Farrella repens, 246, **247**
father lasher, *see Acanthocottus scorpius*
Ficopotamus enigmatica, 78, 102, **103**
fifteen-spined stickleback, *see Spinachia spinachia*
fish, 19, 35, 38–42, 45, 55, 254–71
Fissurellidae, 114
flagellates, 34, 35
flatworms, 27, 34, 55, 70–2
flounder, *see Platichthys flesus*

four-bearded rockling, *see Rhinonemus cimbrius*
Fucus, 24, **26**, 108, 116, 118, 122, 248

Gadidae, 262
Gammarella fucicola, 170, 188, **189**
Gammarellus angulosus, 170, 192, **193**
Gammaridae, 29, **30**, 42, 170, 174–6, 180–7
Gammaropsis maculata, 174, 194, **195**
Gammarus chevreuxi, 175, **183**, 184
Gammarus crinicornis, 175, 184, **186**
Gammarus duebeni, 29, 31, 175, 184, **186**
Gammarus finmarchicus, 175, 184, **186**
Gammarus insensibilis, 29, 31, 51, 175, **183**, 184
Gammarus locusta, 29, 31, 175, 180, **182**
Gammarus oceanicus, 29, 31, 175, 180, **183**
Gammarus pulex, 175, 180, **186**
Gammarus salinus, 29, 31, 176, 180, **183**
Gammarus tigrinus, 175, 176, 184, **186**
Gammarus zaddachi, 29, 31, 176, 180, **183**
garfish, *see Belone bellone*
Garveia franciscana, *see Bimeria franciscana*
Gasterosteidae, 268
Gasterosteus aculeatus, 35, 255, 268, **269**
gastropod molluscs, 20, 25, 33, 35, 43, 54, 55, 105–38, 264, 268
Gastrosaccus sanctus, 157, **159**, 160
Gastrosaccus spinifer, 157, **159**, 160
Gattyana cirrhosa, 80, 86, **87**
Gerridae, 222
Gerris thoracicus, 208, 216, 222, **223**
Gibbula cineraria, 109, 116, **117**
Gibbula umbilicalis, 109, 116, **117**
Gitana sarsi, 172, 178, **179**
Glossiphoniidae, 104
Glycera, 40, **89**
Glycera capitata, 81, 88
Glyceridae, 88
Gnathiidae, 164
Gobiidae, 39, 164, 260, 262, 264
Gobius niger, 257, 264, **265**
goby, *see Pomatoschistus* spp.
godwit, 35
golden (grey) mullet, *see Liza auratus*
Golfingia minuta, *see Nephasoma minuta*
Golfingiidae, 104
Goniodorididae, 134
Goniodoris nodosa, 113, 134, **135**
Gonothyraea loveni, 35, 58, 64, **65**
Grania, 104
Grapsidae, 204
great pipefish, *see Syngnathus acus*
greater waterboatmen, *see Notonecta spp.*

grey topshell, *see Gibbula cineraria*
Gunda ulvae, *see Procerodes littoralis*
Gymnolaemata, 246–50
Gyrinidae, 228
Gyrinus, 209, **229**
Gyrinus caspius, 212, 228, **229**
Gyrinus marinus, 212, 228, **229**

Haemonia mutica, *see Macroplea mutica*
Haliplanella lineata, 59, 66, **67**
Haliplidae, 230
Haliplus, 209, 214, **229**
Haliplus apicalis, 215, **229**, 230
Haliplus confinis, 215, **229**, 230
Haliplus flavicollis, 215, 230, **231**
Haliplus immaculatus, 215, 230, **231**
Haliplus obliquus, 215, **229**, 230
Haliplus wehnckei, 215, **229**, 230
Halitholus cirratus, 58, 60, 62, **63**
Haminea navicula, 106, 130, **131**
Haplotaxida, 104
hard-shelled clam, *see Mercenaria mercenaria*
Harmothoë (Antinoëlla) sarsi, 80, 86, **87**
Harmothoë imbricata, 80, 86, **87**
Harmothoë spinifera, 80, 86, **87**
harpacticoid copepods, 27
Hartlaubella gelatinosa, 58, 64, **67**
Haustoriidae, 188, 190
Haustorius arenarius, 172, 190, **191**
Helcion pellucidum, 107, 114, **115**
Heleobia stagnorum, 112, 120, **121**
Helobdella stagnalis, 102, **103**, 104
Helophorus, 50, 210, **237**
Helophorus aequalis, 213, 236, **237**
Helophorus alternans, 213, 236, **237**
Helophorus arvernicus, 213, 236, **237**
Helophorus brevipalpis, 213, 236, **237**
Helophorus flavipes, 213, 236, **237**
Helophorus minutus, 213, 236, **237**
Hemiptera, 220–3
Hermaea dendritica, 113, **131**, 132
herring, *see Clupea harengus*
Hesperocorixa sahlbergi, 217, 220, **223**
Heteranomia squamula, 140, 146, **147**
Heteroceridae, 242
Heterocerus, 209, **241**
Heterocerus fenestratus, 212, 242, **243**
Heterocerus flexuosus, 212, 242, **243**
Heterocerus maritimus, 212, 242, **243**
Heterocerus obsoletus, 212, 242, **243**
Heteroglossa, 126
Heteromastus filiformis, 38–40, 81, 96, **97**
Heteronemertea, 74
Heteropanope harrisi, *see Rhithropanopeus harrisi*
Heterosomata, 268–70
Heterotanais oerstedi, 168, 169, **179**
Hinia reticulata, 107, 126, **127**
Hinia reticulata var. *nitida*, 126
Hirudinea, 104
Hoplonemertea, 74

houting, *see Coregonus oxyrinchus*
Hyale nilssoni, 173, 180, **181**
Hyalidae, 180
Hydraenidae, 236
Hydridae, 60
Hydrobia, 5, 22, 33, 35, 36, 40, 42–4, 68, 72, 128
Hydrobia acuta, see *Hydrobia neglecta*
Hydrobia jenkinsi, see *Potamopyrgus antipodarum*
Hydrobia minoricensis, see *Hydrobia neglecta*
Hydrobia neglecta, 31, 32, 111, 120, **121**
Hydrobia stagnorum, see *Hydrobia ventrosa* and *Heleobia stagnorum*
Hydrobia truncata, see *Hydrobia ventrosa*
Hydrobia ulvae, 31, 39, 45, **46**, 111, 120, **121**
Hydrobia ventrosa, 31, 111, 120, **121**
Hydrobiidae, 120
Hydrobius fuscipes, 210, 214, 236, **239**
hydroids, 25, 56, 60–5, 112, 134, 136, 169, 180, 194, 198, 200, 246, 248, 250
Hydrophilidae, 236
Hydroporus, 210, **235**
Hydroporus angustatus, 216, 232, **235**
Hydroporus planus, 216, 232, **235**
Hydroporus pubescens, 216, 232, **235**
Hydroporus tessellatus, 216, 232, **235**
hydrozoans, 60–5
Hygrotus inaequalis, 210, 216, 232, **234**

Idotea balthica, 163, 166, **167**
Idotea chelipes, 164, 166, **167**
Idotea granulosa, 163, 166, **167**
Idotea viridis, see *Idotea chelipes*
Idoteidae, 166, 264
insect larvae, 21, 25, 50, 52, 206–10, 222, 268
insects, 19, 25, 35, 53, 206–43
Iravadiidae, 122
Irus irus, 142, 150, **151**
Isaeidae, 194
Ischnochitonida, 105
Ischnochitonidae, 105
Ischnura elegans, 208, 218, **219**
Ischyroceridae, 198–200
Ischyrocerus anguipes, 173, 174, **199**, 200
isopod crustaceans, 25, 35, 45, 52, 162–8, 260, 262, 264
Isospondyli, 258–260

Jaera albifrons, 164, 166
Jaera albifrons agg., 162, 166, **167**
Jaera forsmani, 164, 168
Jaera ischiosetosa, 164, 166
Jaera marina, see *Jaera albifrons* agg.
Jaera nordmanni, 162, 166, **167**
Jaera praehirsuta, 164, 168
Janiridae, 166, 168
Japanese oyster, *see Crassostrea gigas*
Jassa falcata, 173, **199**, 200

jellyfish, *see* scyphozoans

Kellia suborbicularis, 142, 146, **147**
Kellidae, 146
king rag, *see Nereis virens*
knot, 36, 39

Laccobius biguttatus, 210, 214, 236, **238**
Lacuna pallidula, 108, 116, **117**
Lacuna vincta, 110, 118, **119**
Lacunidae, 116, 118
lagoon cockle, *see Cerastoderma glaucum*
lampern, *see Lampetra fluviatilis*
Lampetra fluviatilis, **253**, 254, 258
Lamprothamnium papulosum, 25, **26**, 51
Lanice conchilega, 34, 78, 86, 100, **101**
Laomedea flexuosa, 58, 64, **65**, 136
Laomedea loveni, see *Gonothyraea loveni*
Lasaea rubra, 141, 146, **147**
Lasaeidae, 146
Leander squilla, see *Palaemon elegans*
leeches, 53, 102–4
Lekanesphaera hookeri, see *Sphaeroma hookeri*
Lekanesphaera levii, see *Sphaeroma monodi*
Lekanesphaera rugicauda, see *Sphaeroma rugicauda*
Lembos websteri, 174, 194, **195**
Lepidochitona cinereus, 105, **115**
Lepidonotus squamatus, 80, 86, **87**
Lepidopleurida, 105
Lepidopleuridae, 105
Lepidoptera, 224
Leptoceridae, 224
Leptocheirus pilosus, 172, 194, **195**
Leptochiton asellus, 105, **115**
Leptomysis gracilis, 156, 160, **161**
Leptomysis mediterranea, 156, 160, **161**
Leptoplana tremellaris, 70, 72, **73**
Leptoplanidae, 72
lesser pipefish, *see Syngnathus rostellatus*
lesser waterboatmen, *see Corixidae*
Leucophytia bidentata, 107, 138, **145**
Leucothoe, 172, 180, **181**
Leucothoidae, 180
Libellulidae, 218
Ligia oceanica, 163, **167**, 168
Ligiidae, 168
Liljeborgia kinahani, 173, 190, **191**
Liljeborgiidae, 190
Limanda limanda, 255, 270, **271**
Limapontia capitata, **46**, 114, 132, **133**
Limapontia depressa, 114, 132, **133**
Limapontia senestra, 114, 132, **133**
Limapontiidae, 132
Limnephilidae, 224
Limnephilus affinis, 208, 224, **225**
Limnodrilus, 104
Lineidae, 74
Lineus, 35, **71**
Lineus bilineatus, 71, 74, **75**
Lineus gesserensis, see *Lineus viridis*
Lineus longissimus, 71, 74, **75**

Lineus ruber, 72, 74, **75**
Lineus sanguineus, 71, 74, **75**
Lineus viridis, 72, 74, **75**
Littorina aestuarii, see *Littorina obtusata*
Littorina littorea, 110, 118, **119**
Littorina mariae, 108, 118, **119**
Littorina obtusata, 108, 118, **119**
Littorina saxatilis, 30, 72, 110, 118, **119**
Littorina saxatilis var. *lagunae*, 31, 108, 111, 118, **119**
Littorina saxatilis var. *tenebrosa*, 111, 118, **119**
Littorinidae, 118
Liza auratus, 256, 266, **267**
Liza ramada, 256, 266, **267**
lobworm, *see Arenicola marina*
lugworm, *see Arenicola marina*
Lumbricillus, 104
Lumbriculida, 102–4
Lumbriculidae, 102, **103**, 104
Lumbriculus variegatus, 102
Lymnaea, 31, 106, 138
Lymnaea peregra, 106, 138, **145**
Lymnaeidae, 138
Lysianassa ceratina, 172, 178, **179**
Lysianassidae, 178

Machilidae, 218
Macoma balthica, 30, 34, 36, 41, 43, 143, 150, **151**
Macroplea mutica, 209, 212, 240, **241**
Mactridae, 150
Maera grossimana, 173, 188, **189**
Magelona mirabilis, 81, 94, **95**
Magelonidae, 94
Malacoceros fuliginosus, 40, 82, 92, **93**
malacostracan crustaceans, 156–205
Manayunkia aestuarina, 84, 100, **101**
Marenzelleria viridis, 82, 92, **93**
Margarites helicinus, 109, 116, **117**
Marinogammarus marinus, see *Chaetogammarus marinus*
Marinogammarus obtusatus, see *Eulimnogammarus obtusatus*
Marinogammarus pirloti, see *Chaetogammarus pirloti*
Marinogammarus stoerensis, see *Chaetogammarus stoerensis*
Marionina, 104
Melinna palmata, 84, 98, **99**
Melita palmata, 173, 188, **189**
Melita pellucida, 173, 188, **189**
Melitidae, 188
Membranipora aurita, see *Callopora aurita*
Membranipora crustulenta, see *Electra crustulenta*
Membraniporidae, 248
Mercenaria mercenaria, 142, 148, **149**
Mercierella enigmatica, see *Ficopotamus enigmatica*
Mesopodopsis slabberi, 156, 160, **161**
Metopa pusilla, 172, 180, **181**
Microdeutopus gryllotalpa, 174, 194, **195**

Microprotopus maculatus, 174, 194, **195**
Modiolarca tumida, 141, 144, **145**
Modiolus adriaticus, 141, 144, **145**
Modiolus adriaticus var. *ovalis*, 144
Molgula manhattensis, 252, **253**
Molgulidae, 252
Monia patelliformis, 140, 146, **147**
Monopylephorus, 104
Montacutidae, 146
moon jellyfish, *see Aurelia aurita*
mudsnails, *see Hydrobia*
Mugilidae, 36, 266
Muricidae, 126
Musculus costulatus, 141, 144, **145**
Musculus discors, 141, 144, **145**
mussels, *see Mytilus edulis*
Mya, 40, 264
Mya arenaria, 43, 139, 152, **153**
Mya truncata, 140, 152, **153**
Myacidae, 152
Myida, 152
Myoxocephalus scorpius, see Acanthocottus scorpius
Mysella bidentata, 142, 146, **147**
mysid crustaceans, 19, 53, 156–61, 260
Mysidae, 158–60
Mysta picta, see Eteone picta
Mystacides longicornis, 208, 224, **225**
Mytilida, 144
Mytilidae, 144
Mytilopsis leucophaeta, see Congeria cochleata
Mytilus edulis, 22, 34, 43, 128, 141, 144, **145**
Myxicola infundibulum, 84, 100, **101**

Naididae, 102, **103**, 104
Nais, 104
Najas marina, 25
Nassariidae, 126
Nassarius reticulata, see Hinia reticulata
native oyster, *see Ostrea edulis*
Neanuridae, 218
nematodes, 27, 34, 55, 60, 76, **77**
Nematostella vectensis, 27, 35, 40, 51, 58, 68, **69**
nemerteans, *see nemertines*
nemertines, 35, 55, 70–6
Neomysis integer, 156, 158, **159**
Neomysis vulgaris, see Neomysis integer
Neotaenioglossa, 116–26
Nephasoma minuta, 22, **103**, 104
Nephtys, 40, **79**, 81, 85
Nephtys caeca, 85, 90, **91**
Nephtys cirrosa, 85, 90, **91**
Nephtys hombergi, 85, 90, **91**
Nephtys longosetosa, 85, 90, **91**
Nereidae, 90
Nereis, 40, 68, 146, 260
Nereis (Hediste) diversicolor, 5, 23, **24**, 30, 33–5, 44, 45, **46**, 83, 90, **91**
Nereis (Neanthes) succinea, 83, 90, **91**

Nereis (Neanthes) virens, 83, 90, **91**
Neritida, 116
Neritidae, 116
Nerophis ophidion, 254, 262, **263**
Noteridae, 230
Noterus clavicornis, 209, 216, 230, **231**
Notodorididae, 136
Notonecta glauca, 216, 222, **223**
Notonecta viridis, 216, 222, **223**
Notonectidae, 31, 208, 222
Nucella lapillus, 107, 126, **127**
nudibranchs, 134–7
Nymphula nymphaeata, 208, 224, **225**

Obelia bidentata, 58, 64, **67**
Obelia dichotoma, 58, 64, **65**
Obelia gelatinosa, see Hartlaubella gelatinosa
Obelia longissima, 58, 64, **65**
Ochthebius, 50, **239**
Ochthebius aeneus, 213, 236, **239**
Ochthebius auriculatus, 214, 236, **239**
Ochthebius dilatatus, 214, 236, **239**
Ochthebius exaratus, 213, 236, **239**
Ochthebius lenensis, 213, 236, **239**
Ochthebius marinus, 213, 236, **239**
Ochthebius nanus, 214, 236, **239**
Ochthebius punctatus, 214, 236, **239**
Ochthebius quadricollis, 236
Ochthebius viridis, 213, 236, **239**
Odonata, 218
Oecetis ochracea, 208, 224, **225**
Oedicerotidae, 190
Oerstedia dorsalis, 70, 76, **77**
oligochaetes, 31, 49, 50, 53, 102–4, 264
Oligoentomata, 218
Omalogyra atomus, 108, 126, **127**
Omalogyridae, 126
Onchidorididae, 134
Onchidoris muricata, 113, 134, **135**
Onoba aculeus, 22, 109, 110, 122, **123**
Onoba semicostata, 109, 122, **123**
Opheliida, 98
Opheliidae, 98
Ophelia bicornis, 81, 98, **99**
Ophelia rathkei, 81, 98, **99**
Ophiuroidea, 55, 250
opisthobranch molluscs, 35, 55, 112–14, 128–137
oppossum shrimps, *see mysid crustaceans*
Orbiniida, 92
Orbiniidae, 92
Orchestia, 173, 180, **181**
Orchestia remyi roffensis, 180
Orthetrum cancellatum, 208, 218, **219**
Orthocladiinae, 50
Osmeridae, 260
Osmerus eperlanus, 256, 260, **261**
Osteichthyes, 258–71
ostracod crustaceans, 27, 54, **155**, 156
Ostrea edulis, 140, 144, **145**

Ostreidae, 144
Ovatella myosotis, 107, 138, **145**
oystercatcher, 35, 36

Pacific oyster, *see Crassostrea gigas*
Palaemon adspersus, 201, 202, **203**
Palaemon elegans, 201, 202, **203**
Palaemon longirostris, 201, 202, **203**
Palaemon serratus, 201, 202, **203**
Palaemon squilla, see Palaemon elegans
Palaemonetes varians, 35, 40, 201, 202, **203**
Palaemonidae, 202
Palaeonemertea, 72
Paludinella littorina, 22, 51, 108, 124, **125**
Pandeidae, 62
Paracorixa concinna, 217, 220, **221**
Paracymus aeneus, 51, 210, 214, 236, **238**
Paragnathia formica, 162, 164, **165**
Paramysis nouveli, 158, **159**
Paranais, 104
Paraonidae, 92
Paraonis fulgens, 80, 92, **93**
Paratanaidae, 169
Parvicardium exiguum, 140, 141, 148, **149**
Parvicardium hauniense, 140, 148, **149**
Pasiphaea sivado, 200, 202, **203**
Pasiphaeidae, 202
Patella vulgata, 107, 114, **115**
Patellidae, 114
Pectenogammarus planicrurus, 174, 185, **187**
Peloscolex, see Tubificoides
Percomorphi, 264, 266
Perigonimus cirratus, see Halitholus cirratus
Perigonimus megas, see Bimeria franciscana
Perinereis cultrifera, 83, 90, **91**
Petrobius brevistylis, 208, 218, **219**
Petromyzon marinus, **253**, 254, 258
Petromyzonidae, 258
Pholadidae, 152
Pholoë minuta, 80, 86, **87**
Phoxocephalidae, 190
Phoxocephalus holbolli, 172, 190, **191**
Phragmites, 15, 24, 25, 27, 35, 62, 102, 202, 230, 246, 248
Phylactolaemata, 246
Phyllodoce groenlandica, 84, 88, **89**
Phyllodoce maculata, 84, 88, **89**
Phyllodocida, 86–91
Phyllodocidae, 88
Phytia myosotis, see Ovatella myosotis
pilchard, *see Sardina pilchardus*
Pilumnopeus harrisi, see Rhithropanopeus harrisi
Pilumnus tridentatus, see Rhithropanopeus harrisi
plaice, *see Pleuronectes platessa*
planarians, *see triclads*

Platichthys flesus, 36, 37, 164, 255, 270, **271**
platyhelminths, 72
Pleurobranchidae, 130
Pleurobranchomorpha, 130
Pleurogona, 252
Pleuronectes platessa, 255, 270, **271**
Pleuronectidae, 270
Pleurotomariida, 114
plovers, 35
Plumatella fungosa, 245, 246, **247**
Plumatella repens, 245, 246, **247**
Plumatellida, 246
Plumatellidae, 246
Plumulariidae, 64
pogge, *see Agonus cataphractus*
polychaetes, 34, 35, 43, 53, 78–102, 270
polyclads, *see Polycladida*
Polycladida, 72
Polydora, 40
Polydora (Boccardia) ligerica, 82, 94, **95**
Polydora (Boccardia) redeki, see Polydora (Boccardia) ligerica
Polydora (Polydora) ciliata, 81, 94, **95**
Polydora (Polydora) ligni, 82, 94, **95**
Polydora (Pseudopolydora) pulchra, 82, 94, **95**
Polynoidae, 86
polyplacophoran molluscs, *see* chitons
Pomatoschistus microps, 27, 35, 148, 257, 264, **265**
Pomatoschistus minutus, 257, 264, **265**
pondweeds, *see Potamogeton* spp.
Pontocrates, 172, 190, **191**
Pontoporeia femorata, 170, 190, **191**
Portunidae, 204
pot worms, *see* Enchytraeidae
Potamogeton filiformis, 25
Potamogeton pectinatus, 25, **26**, 148, 220, 224, 236, 240
Potamonectes depressus, 210, 216, 232, **235**
Potamopyrgus antipodarum, 109, 111, 120, **121**
Potamopyrgus jenkinsi, see Potamopyrgus antipodarum
Potamothrix, 104
pout, *see Trisopterus luscus*
praniza larvae, 164
Praunus flexuosus, 158, **159**
Praunus inermis, 158, **159**
prawns, 19, 25, 39, 200–5, 262
Priapus equinus, see Actinia equina
Procerodes littoralis, 70, 72, **73**
Procerodidae, 72
prosobranchs, 114–28
Prosorhochmidae, 76
Prosorhochmus claparedii, 22, 70, 76, **77**
Prostomatella obscurum, 70, 76, **77**
Protohydra leuckarti, 35, 55, 59, 60, **61**, 136
Psammodrilida, 96

Psammodrilidae, 96
Psammodrilus balanoglossoides, 80, 96, **97**
Pseudamnicola confusa, 51, 111, 120, **121**
Psychodidae, 50, 206, 226, **227**
Pteriida, 114
Pterygota, 218–43
Ptychopteridae, 50, 206, 226, **227**
pulmonate molluscs, 138
Pungitius pungitius, 255, 268, **269**
purple topshell, *see Gibbula umbilicalis*
Pusillina inconspicua, 109, 122, **123**
Pusillina sarsi, 109, 111, 122, **123**
Pygospio elegans, 40, 43, **46**, 82, 94, **95**
Pyralidae, 208, 224
Pyramidellidae, 128
Pyramidellomorpha, 128

ragworm, *see Nereis diversicolor* and *Perinereis cultrifera*
Rathkea octopunctata, 60, 62, **63**
Rathkeidae, 62
razor-shell, *see Ensis directus*
redshank, 35, 36, 38
reeds, *see Phragmites*
Retusa obtusa, 106, 128, **129**
Retusa truncatula, 106, 128, **129**
Retusidae, 128
Rhantus exsoletus, 209, 215, 232, **233**
Rhinonemus cimbrius, 257, 262, **263**
Rhithropanopeus harrisi, 200, 204, **205**
Rhizorhagium navis, see Clavopsella navis
Rhynchobdellida, 104
ribbonworms, *see* nemertines
Rissoa albella, see Pusillina sarsi
Rissoa parva, 109, 122, **123**
Rissoa parva var. *interrupta*, 111
Rissoa rufilabrum, 109, 122, **123**
Rissoella diaphana, 111, 124, **125**
Rissoella globularis, 108, 124, **125**
Rissoella opalina, 110, 124, **125**
Rissoellidae, 124
Rissoidae, 122–3
Rissostomia labiosa, 107, 109, 110, 124
Rissostomia membranacea, 45, 72, 107, 109, 110, 124, **125**
rough winkle, *see Littorina saxatilis*
roundworms, *see* nematodes
ruddy darter, *see Sympetrum sanguineum*
Runcina coronata, 114, 130, **131**
Runcinidae, 130
Ruppia, 35, 40, 64, 68, 92, 94, 100, 108, 118, 126, 132, 148, 164, 166, 192, 194, 202, 220, 224, 226, 236, 240, 246, 248, 268
Ruppia cirrhosa, 25, **26**
Ruppia maritima, 25, **26**

Sabella pavonina, 38, 84, 100, **101**
Sabella penicillus, see Sabella pavonina
Sabellaria alveolata, 78, 86, 98, **99**, 104
Sabellaria spinulosa, 78, 86, 98, **99**, 104
Sabellariidae, 98
Sabellida, 100–2

Sabellidae, 100
Sacoglossa, 132
saddle-oysters, *see* Anomiidae
Sagartia elegans, 59, 68, **69**
Sagartia ornata, 59, 68, **69**
Sagartia troglodytes, 59, 68, **69**
Sagartiidae, 68, 136
Saldidae, 222
Saldula palustris, 209, 212, 222, **223**
Salmo salar, 256, 260, **261**
Salmo trutta, 256, 260, **261**
salmon, *see Salmo salar*
Salmonidae, 260
sand mason, *see Lanice conchilega*
sand-gaper, *see Mya arenaria*
sand-smelt, *see Atherina presbyter*
Sardina pilchardus, 257, 258, **259**
sardine, *see Sardina pilchardus*
Sarsia loveni, 56, 60, **61**
Sarsia tubulosa, 56, 60, **61**
scaleworms, *see* Polynoidae
scallops, 128
scavenger beetles, *see* Hydrophilidae
Schistomysis kervillei, 158, 160, **161**
Schistomysis ornata, 158, 160, **161**
Schistomysis parkeri, 157, 160, **161**
Schistomysis spiritus, 158, 160, **161**
Scleroparei, 266, 268
Scolecolepides viridis, 92
Scolelepis foliosa, 82, 92, **93**
Scolelepis squamata, 82, 92, **93**
Scoloplos armiger, 81, 92, **93**
Scophthalmus maximus, 255, 268, **269**
Scophthalmus rhombus, 255, 268, **269**
Scrobicularia plana, 23, **24**, 34, 143, 150, **151**
Scrobiculariidae, 150
scyphozoans, 56, 66
sea anemones, 21, 22, 25, 56, 112
sea lamprey, *see Petromyzon marinus*
sea mouse, *see Aphrodita aculeata*
sea scorpion, *see Acanthocottus scorpius*
sea trout, *see Salmo trutta*
sea-grasses, *see Zostera*
sea-slugs, *see* opisthobranch molluscs
sea-squirts, *see* ascidians
seaweeds, 158
see also algae
Semibalanus balanoides, 154, **155**
Semisalsa stagnorum, see Heleobia stagnorum
Serpulidae, 102
Serranidae, 264
Sertularia cupressina, 58, 64, **65**, 134
Sertulariidae, 64
shelduck, 35
shore crab, *see Carcinus maenas*
shrimps, *see Crangon crangon*
Sigalionidae, 86
Sigara dorsalis, 217, 220, **221**
Sigara falleni, 217, 220, **221**
Sigara lateralis, 217, 220, **221**
Sigara selecta, 217, 220, **221**

Sigara stagnalis, 217, 220, **221**
Sigara striata, 51, 217, 220, **221**
Siphonostoma typhle, 254, 262, **263**
sipunculans, 55, 104
Skeneopsidae, 126
Skeneopsis planorbis, 108, 126, **127**
smelt, *see Osmerus eperlanus*
soft-shelled clam, *see Mya arenaria*
Solea solea, 254, 270, **271**
Soleidae, 270
solenette, *see Buglossidium luteum*
Solenichthyes, 262
Solenidae, 150
Sphaeroma hookeri, 163, 164, **165**
Sphaeroma monodi, 163, **165**, 166
Sphaeroma rugicauda, 163, **165**, 166
Sphaeromatidae, 164–5
Spinachia spinachia, 255, 268, **269**
Spio filicornis, 40, 82, 94, **95**
Spio martinensis, see Spio filicornis
Spionida, 92–96
Spionidae, 34, 78, 88, 92–5
Spiophanes bombyx, 82, 92, **93**
Spirorbidae, 78, 102, **103**
Spisula subtruncata, 40, 142, 150, **151**
sprat, *see Sprattus sprattus*
Sprattus sprattus, 257, 258, **259**
Staphylinidae, 240, **241**
Stenoglossa, 126
Stenothoe monoculoides, 172, 180, **181**
Stenothoidae, 180
stickleback, *see Gasterosteus aculeatus,*
    Pungitius pungitius, Spinachia
    spinachia
Stiliger bellulus, 113, 132, **133**
Stiligeridae, 132
Stratiomyiidae, 50, 206, 226, **227**
Streblospio shrubsoli, 82, 92, **93**
Streptosyllis websteri, 83, 88, **89**
Styela clava, 252, **253**
Styelidae, 252
Stylaria, 104
Syllidae, 88
Sympetrum sanguineum, 208, 218, **219**
Synentognathi, 260
Syngnathidae, 262

Syngnathus acus, 254, 262, **263**
Syngnathus rostellatus, 254, 262, **263**
Syrphidae, 50, 206, 226, **227**

Tabanidae, 50, 207, 226, **227**
Talitridae, 180
Talitrus saltator, 173, 180, **181**
tanaid crustaceans, 43, 53, 168, 169
Tanaidae, 168
Tanais dulongii, 168, **179**
Tealia felina, see Urticina felina
Tellinidae, 150
ten-spined stickleback, *see Pungitius*
    pungitius
Tenellia adspersa, 35, 51, 113, 136, **137**
Terebellida, 98–100
Terebellidae, 100
Terebellides stroemi, 84, 100, **101**
Tergipedidae, 136
Tetrastemma melanocephalum, 70, 76, **77**
Tetrastemmatidae, 76
Tharyx marioni, 40, 81, 96, **97**
thecate hydroids, 64, 65, 134
Theodoxus fluviatilis, 108, 116, **117**
thick-lipped grey mullet, *see Crenimugil*
    labrosus
thin-lipped grey mullet, *see Liza*
    ramada
Thoracica, 154
three-spined stickleback, *see*
    Gasterosteus aculeatus
Thysanura, 218
Tipulidae, 50, 206, 226, **227**
Trichobranchidae, 100
Trichoptera, 224
triclads, 72
Trisopterus luscus, 257, 262, **263**
Triticellidae, 246
Trochidae, 116
Truncatella subcylindrica, 22, 51, 109,
    122, **123**, 124
Truncatellidae, 122
Tubifex, 104
Tubificidae, 49, 102, **103**, 104
Tubificoides, 40, 104
Tubularia indivisa, 56, 60, **61**

Tubularia larynx, 56, 60, **61**
Tubulariidae, 60
Tunicata, 252
turbellarians, 72
Turbicellepora avicularis, 245, 250, **251**
Turbonilla lactea, 108, 128, **129**
turbot, *see Scophthalmus maximus*
twaite shad, *see Alosa fallax*

Ulva, 24, 33, 126, 158
Uncinais uncinata, 104
Urothoe, 172, 190, **191**
Urticina felina, 59, 66, **67**
Uteriporidae, 72
Uteriporus vulgaris, 70, 72, **73**

Venerida, 146
Veneridae, 148
Venerupis pullastra, see Venerupis
    senegalensis
Venerupis senegalensis, 142, 148, **149**
Ventromma halecioides, 58, 64, **65**
Vesiculariidae, 248
Victorella pavida, 31, 51, 245, 246, **247**
Victorellidae, 246

whelk, *see Buccinum undatum*
whirligig beetles, *see Gyrinus* spp.
whitebait, 258, 260
winkles, *see Littorinidae*
worm pipefish, *see Nerophis ophidion*
wracks, *see Fucus*

Xanthidae, 204

Zannichellia palustris, 25
Zoarces viviparus, 257, 264, **265**
Zoarcidae, 264
Zostera, 24, **26**, 64, 68, 72, 74, 76, 90,
    92, 94, 96, 98, 100, 116, 118,
    122, 124, 126, 130, 132, 144,
    148, 160, 192, 194, 198, 202,
    224, 240, 260, 262, 264, 268
Zygoentomata, 218